I0814314

# Diamonds

# Diamonds

## Their History, Sources, Qualities and Benefits

**Renée Newman, GG**

Firefly Books

Published by Firefly Books Ltd. 2021


First printing

**Library of Congress Control Number: 2021936648**

**Library and Archives Canada Cataloguing in Publication**
Title: Diamonds : their history, sources, qualities and benefits / Renée Newman, GG.
Names: Newman, Renée, 1948- author.
Description: Includes bibliographical references and index.
Identifiers: Canadiana 20210185813 | ISBN 9780228103318 (hardcover)
Subjects: LCSH: Diamonds. | LCSH: Diamond industry and trade.
Classification: LCC QE393 .N49 2021 | DDC 553.8/2—dc23

Published in the United States by
Firefly Books (U.S.) Inc.
P.O. Box 1338, Ellicott Station
Buffalo, New York 14205

Published in Canada by
Firefly Books Ltd.
50 Staples Avenue, Unit 1
Richmond Hill, Ontario L4B 0A7

Cover and interior design: Hartley Millson
Editor: Julie Takasaki
Copy editor: Ronnie Shuker
Indexer: Emmalena Petersen
Proofreader: Nancy Foran

Front cover image: Rings and photo courtesy of Dehres Ltd.
Back cover images: Diamonds and photos courtesy of Dehres Ltd. and Joe Namdar
Page 2: Necklace and photo courtesy of Dehres Ltd.
Page 5: Pendant by Paula Crevoshay; photo by Crevoshay Studio

Printed in China

# Contents

## 1 What Is a Diamond? . . . 10



## 2 Where Are Diamonds Found? . . . 38

## 3 Diamond Mining and Processing . . . 78



## 4 The Evolution of Diamond Cutting . . . 96

# 5 The Evolution of Diamond Jewelry . . . 132



# 6 How Are Diamonds Priced? . . . 184

# 1 What Is a Diamond?

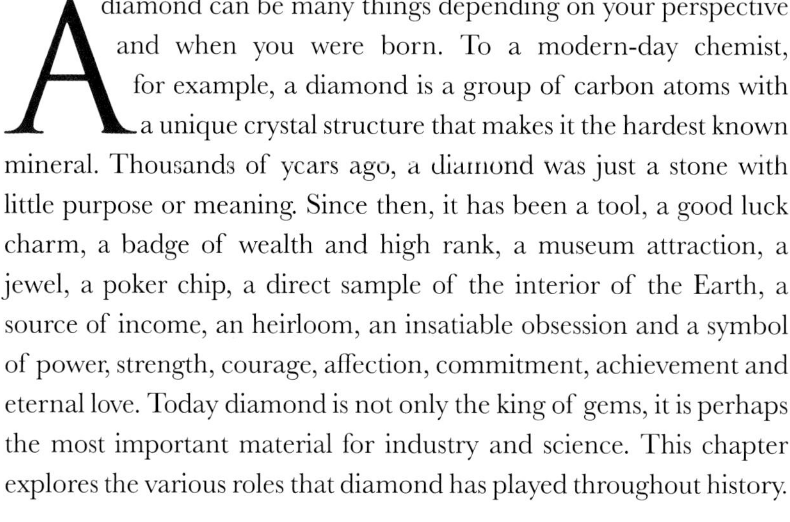

A diamond can be many things depending on your perspective and when you were born. To a modern-day chemist, for example, a diamond is a group of carbon atoms with a unique crystal structure that makes it the hardest known mineral. Thousands of years ago, a diamond was just a stone with little purpose or meaning. Since then, it has been a tool, a good luck charm, a badge of wealth and high rank, a museum attraction, a jewel, a poker chip, a direct sample of the interior of the Earth, a source of income, an heirloom, an insatiable obsession and a symbol of power, strength, courage, affection, commitment, achievement and eternal love. Today diamond is not only the king of gems, it is perhaps the most important material for industry and science. This chapter explores the various roles that diamond has played throughout history.

## A Practical Tool

A portrait of Pliny the Elder. *Science History Images/Alamy Stock Photo*

The first role that diamond had was as a tool for cutting, drilling, grinding and engraving. Ancient civilizations in India, Greece and Rome gradually noticed that diamond was harder than any other material, which in turn allowed it to be used for practical purposes. Archaeological evidence suggests that diamond splinters were used to drill beads in India as early as the fifth century BCE.

The name diamond derives from the Greek word *adamas*, meaning "unconquerable force" — in reference to its extreme hardness. Despite its resistance to scratching, abrasion and deformation, diamond can be split if hit in the right spot. In 77 CE, the Roman author Pliny the Elder wrote in Book 37 of his *Natural History*, "When an

A diamond-tipped drill in action. Diamonds continue to have vast industrial and practical applications. *David Tadevosian/Shutterstock*

'adamas' is successfully broken, it disintegrates into splinters so small as to be scarcely visible. These are much sought after by engravers of gems and are inserted by them into iron tools because they make hollows in the hardest materials without difficulty."

In India, Hindu artisans noticed that when a diamond was struck on an anvil it did not smash but embedded itself in the anvil. However, they also discovered that under some conditions, an anvil blow could break a diamond, so they began to wrap the diamonds in sheets of lead or wax and hit them sharply. Then they opened the sheets, lined up the splinters and struck them with the edge of a heated sword or the tip of a hot tool to make diamond-edged knives and swords and diamond-tipped tools.

The Chinese knew diamond first as a "jade-cutting knife," not as a jewel. Jade was their most honored gem. Since diamond was so often found near gold, they thought it was related to gold.

In the 1300s Indians and Europeans began to shape diamonds by using a surface coated with diamond grit and olive oil to polish off bumps and depressions on rough crystals. By the late 1800s diamond polishing had been mechanized but still involved rubbing two diamond crystals against each other, one of which was the tool and the other the gem.

Before World War II, diamonds were used in a wide variety of industrial applications, including drills, saws, excavation, aircraft systems, phonograph needles and surgical blades. When the war began, diamond cutting became even more important for creating weapons, vehicles and technology. The Germans put imprisoned diamond cutters in special camps to work on war machines, and they scheduled raids to obtain inventory. When the Russians invaded Germany, they abducted cutters in the town of Idar-Oberstein and placed them in forced labor camps.

Diamond's importance as a tool today extends beyond cutting and drilling. It has many extraordinary physical and optical properties that make it useful for other applications in industry, scientific exploration and medical technology. These are discussed further in chapter 8.

## A Good Luck Charm with Magical Powers

In ancient India, diamonds were assumed to have supernatural powers because no other stone except another diamond could scratch or deform it due to its superior hardness. As a result, diamonds were worn to give the owner strength, courage and invincibility in battle. Diamond was always considered a stone of winners, provided it was of good quality and did not have cracks or other defects; it was the talisman of both Julius Caesar and Napoleon.

It was also believed that diamonds had healing properties and could provide protection from illness. A supposed proof of this was that the Black Death, a plague that swept through Europe, Asia and Africa in the 1300s, first attacked the poorer classes, largely sparing the rich, who could afford to wear diamonds.

According to the book *The Curious Lore of Precious Stones* by George F. Kunz (1856–1932), the Hindus classed diamonds according to the four castes. The Brahmin diamond gave power, friends, riches and good luck; the Kshatriya diamond prevented the approach of old age; the Vaishya stone brought success; and the Shudra diamond brought all manner of good fortune.

Even today, authors of metaphysical books attribute powers to diamonds. In *The Crystal Bible*, author Judy Hall says "Psychologically the qualities that diamond imparts include fearlessness, invincibility, and fortitude … Diamond treats glaucoma, clears sight and benefits the brain. It treats allergies and chronic conditions and rebalances the metabolism. Traditionally it was used to counteract poisons."

In *The Book of Stones*, authors Robert Simmons and Naisha Ahsian discuss how many monarchs had diamonds set in crowns because they believed diamonds provided access to divine energies: "Putting such a stone near the brain, particularly the forehead, can enhance one's inner vision and intuitive connection with the higher domains of the spirit … Diamond is a tool one can use to evoke the inner king and/or queen, those archetypal beings within the self that convey power, knowledge and sovereignty."

The Koh-i-noor diamond currently sits front and center in the Queen Mother's coronation crown. *World History Archive/Alamy Stock Photo*

## A Symbol of Power and Wealth

In India large diamonds were badges of rank and wealth worn by rulers. This tradition spread throughout Europe, with the wealthiest and most powerful families using diamonds and other gems to solidify their status as rulers of the world.

One of the biggest and most famous diamonds is the 186-carat Koh-i-noor, which means "mountain of light" in Persian. The diamond, which was found in India, dates back to at least 1304 and had changed hands among various rulers in India, Iran, Afghanistan and Pakistan when the East India Company took it over in 1849 for Queen Victoria. It was once said that whoever owned the Koh-i-noor ruled the world. One of its owners, Sultan Babur, referred to it in his diary in 1526 as "the famous diamond of such value that it would pay half the expenses of the world."

In 1850 the Koh-i-noor was presented to Queen Victoria, and the next year the 186-carat diamond was displayed at the Great Exhibition in London. There were complaints that its Indian-style cut made it look dull, so Prince Albert, the husband of Queen Victoria, ordered it recut as an oval brilliant to give it more life. A large yellow flaw was removed in the process, reducing the size of the diamond to 108.93 carats. The cutting of the Koh-i-noor was done in London by cutters from Amsterdam and lasted for 38 days. Afterward it was mounted in a brooch and circlet worn by the queen, but following her death it was set in the crown of Queen Alexandra, the wife of King Edward VII. The Koh-i-noor was transferred to Queen Mary's crown in 1911 and finally to the crown of Queen Elizabeth, wife of King George VI and later known as the Queen Mother, in 1937. All of these crowns are on display at the Jewel House in the Tower of London.

The Koh-i-noor has only been worn by female members of the British royal family because it is believed to bring bad luck to the men who wear it due to the history of misfortunes that befell many of its owners. Some were tortured and died prematurely. However, a few became great conquerors.

There has been much controversy about the rightful owner of the Koh-i-noor. The governments of India, Pakistan, Iran and Afghanistan have all claimed rightful ownership of the diamond and demanded its return ever since India gained independence from the United Kingdom in 1947. The British government insists the Koh-i-noor was obtained legally under the terms of the Last Treaty of Lahore and has rejected the claims. However, one Indian statesman disagreed with his country's claim to the diamond —Jawaharlal Nehru (1889–1964), an Indian independence activist and the first prime minister of independent India. He said, "Diamonds are for the emperors and India does not need emperors."

French kings also displayed their wealth with diamonds. In the 16th and 17th centuries, France led the European diamond market. Louis XIII was a diamond aficionado and a patron of diamond cutting. He is credited with commissioning preliminary models of the brilliant cut. His collection reportedly contained 18 of the largest and finest diamonds in Europe.

Under Louis XIV, the French love affair with diamonds reached its peak. Louis bought 44 large diamonds and more than 1,100 smaller ones from the famous gem merchant Jean-Baptiste Tavernier, including the Tavernier Blue, the diamond that would become the French Blue and later would be named the Hope Diamond.

The Hope Diamond is a major attraction at the Smithsonian Museum of Natural History. *Photo by Chip Clark; courtesy of the Smithsonian Institution*

A painting of the insignia of the Order of the Golden Fleece, the ornament commissioned by Louis XV. The painting shows the French Blue, the diamond that was stolen during the French Revolution and was later recut into the Hope Diamond. Above it is the red Côte de Bretagne spinel (dragon). *Gouache painting created by Pascal Monney, Geneva, Switzerland; reprinted with permission of the owner, Herbert Horovitz*

Socialite Evalyn Walsh McLean wearing the Hope Diamond. *Everett Collection Historical/Alamy Stock Photo*

The Tavernier Blue likely came from the Kollur mine in India. Louis XIV purchased it from Tavernier in 1668, and four years later he had it recut from an Indian cut weighing about 110 carats into a 69-carat heart-shaped stone known as the French Blue.

In 1749 Louis XV asked the crown jeweler to mount the diamond in the insignia of the Order of the Golden Fleece. This ornament was stolen in 1792 during the French Revolution. Around 1812 a dark blue diamond mysteriously appeared in London, England. It had been recut into a 45.52-carat diamond now called the Hope Diamond, which acquired its name when Henry Philip Hope, a member of the banking family Hope & Co., bought it. After going through numerous owners, it was sold to Washington, D.C., socialite Evalyn Walsh McLean for $180,000. New York diamond dealer Harry Winston purchased her entire collection of jewelry from her estate in 1949 and toured with the diamond for several years. In 1958 he presented the Hope Diamond to the Smithsonian Institution, saying he wanted it to be the beginning of a national collection of gems that would rival the one in the Tower of London.

The Regent Diamond. *Photo © RMN-Grand Palais/Art Resource, NY*

Another famous diamond worn by French kings was the Regent Diamond from the Golconda area of India. The diamond weighed about 410 to 426 carats before its owner, Sir Thomas Pitt, had it shipped from India to England in 1702. The diamond had flaws that extended quite deep into the stone, so it was recut between 1704 and 1706 to turn it into an internally flawless D-color (completely colorless) cushion-shaped brilliant-cut diamond, which ended up weighing 140.64 carats. Its blue fluorescence gave it a desirable bluish tinge. During the cutting process, several secondary stones were produced and later sold to Peter the Great, the czar of Russia from 1682 to 1725.

Pitt was constantly worried about losing the diamond or being robbed of it, so he decided to sell it. In 1717 the diamond was purchased for the French Crown at the request of Philippe d'Orléans, the regent of France from 1715 until 1723, when Louis XV became old enough to assume his duties as king. The diamond, which was considered then to be the finest and largest of all known diamonds in the western world, was named after the regent and is still known as "Le Régent" in French and "the Regent" in English.

Louis XV first wore the Regent in 1721 at a reception for Turkish ambassadors at Versailles. It was set in a knot of pearls and diamonds on a shoulder ornament. The Regent remained part of the French crown jewels until it was stolen in 1792. The following year, the diamond was found hidden in some roof timbers of a Paris attic, and then later it was used to secure loans to pay expenses for the French military before Napoleon Bonaparte claimed it in 1801. Afterward Napoleon had the Regent set in the hilt of his sword, which he carried at his coronation as emperor of France. Following changes in the ruling regime, the diamond was mounted successively on the crowns of Louis XVIII, Charles X and Napoleon III and finally on the Grecian diadem (crown) of Empress Eugénie.

The Regent Diamond has belonged to the Louvre Museum in Paris since 1887 and has been on display continuously since then, except when the Germans invaded France in World War II. It was sent to the village of Chambord and hidden in a château for safe keeping until the end of the war. The Regent was then returned to Paris and put on display at the Louvre, where it remains today. Because of its history and fame as a diamond, the Regent has been called the National Diamond of France.

Russia's czars also loved diamonds and chose them as symbols of power. Even in the early 1600s their treasures included a diamond-studded royal orb and diamond-encrusted crowns. Catherine the Great (1729–1796) commissioned Jérémie Pauzié, a Genevan artist and diamond jeweler, to design a crown set with almost 5,000 diamonds. He described it as the richest object that ever existed in Europe.

## An Ornamental Gemstone

Uncut diamond crystals were set in gold rings as early as 300 BCE, but they were generally worn as talismans rather than as jewelry and were reserved for the ruling class. It was only after cutters developed ways to polish diamonds that they became coveted as gems. Prior to the 15th century, pearls, rubies and sapphires were more highly regarded.

Agnès Sorel (1422–1450), the mistress of King Charles VII of France, played a major role in making the diamond esteemed for

**(opposite page)** A replica of the Great Imperial Crown created for the coronation of Catherine the Great in 1762. *Album/Alamy Stock Photo*

*Lanté del.t* *Gatine sculp.t*

AGNÈS SOREL,

*Née vers 1409, Morte en 1450.*

feminine adornment. She is credited with being the first woman and first commoner to wear diamonds in public. Her position as the king's mistress in the 1440s placed her above the laws that forbade the wearing of jewelry except by nobles and clergy.

A French diamond merchant and friend of hers named Jacques Cœur supplied her with diamond necklaces, brooches and buckles. He brought diamond workers from Venice and Constantinople (modern-day Istanbul, Turkey), and he sent traders into India not only for big stones of status but also smaller ones for necklaces and brooches. This influx of gems made it possible for commoners to wear diamonds and helped turn them into a feminine ornament.

Agnès became the model for all the jewels and silken robes that Jacques imported. After seeing diamond jewelry on Agnès, other women were inspired to wear it too. This helped bring business and fashion to Paris. By the 16th and 17th centuries France led the European diamond market.

In the 20th century, Hollywood actresses such as Elizabeth Taylor drew attention to the appeal of diamonds, especially large ones. Taylor's first major diamond was the 33.19-carat Krupp Diamond, which she wore in a ring. But she is more noted for wearing the 69.42-carat Taylor-Burton Diamond, which was the centerpiece of a diamond necklace.

The original rough (uncut diamond) for the Taylor-Burton Diamond weighed 241 carats and was found in 1966 at the Premier mine in South Africa. Harry Winston had it cut into the pear shape weighing 69.42 carats. The diamond was first purchased by Harriet Annenberg Ames, the sister of billionaire publisher Walter Annenberg, but she decided to sell the stone at auction in 1969. Taylor's husband, actor Richard Burton, was one of the bidders, but he lost out to Cartier Jewelers, who bought the diamond for $1,050,000, at the time a record price for a gem sold at auction. The previous record for a diamond was $305,000, set in 1957.

Burton was upset that he had been outbid, and the next day he negotiated the purchase of the diamond from Cartier for $1,100,000 from a hotel pay phone. A condition of the sale was that Cartier would be able to display the diamond in New York and Chicago. An

(opposite page) Agnès Sorel, known as the Dame de Beauté of 15th-century France, was the first commoner to wear diamonds in public. *Album/ Alamy Stock Photo*

Elizabeth Taylor wearing the Taylor-Burton Diamond at the 42nd Academy Awards in April 1970. *Photo by Ron Galella/Ron Galella Collection via Getty Images*

estimated 6,000 people lined up to see it. Taylor wore the diamond to the 42nd Academy Awards, where she presented the award for Best Picture to *Midnight Cowboy*.

After her second divorce from Burton in 1976, Taylor announced that she was putting the diamond up for sale and was planning to use part of the proceeds to build a hospital in Botswana. She sold the Taylor-Burton Diamond to Henry Lambert, a jeweler in New York, who in turn sold it to Robert Mouawad of the Mouawad jewelry house in 1979.

## A Poker Chip or Plaything

In 1726 gold prospectors in the Brazilian state of Minas Gerais would play poker after panning in a stream all day for gold. They used pebbles found in their pans with the gold as poker chips. Unbeknownst to them, many of the stones were diamonds. A Portuguese soldier who had been lured to Brazil by the report of gold and who had seen diamond crystals in India suspected they were diamonds, so he sent a few to Lisbon. Some of the samples were cut in Amsterdam and were determined to be diamonds as good as those found in India. After the King of Portugal learned that diamonds had been found in Brazil, he removed the gold prospectors and placed the fields in the hands of a few court favorites, who then controlled the slave labor gangs and the diamonds. By 1730 Brazil had replaced India as the world's main source of diamonds and kept that position until 1870, when production started in South Africa.

Around late 1866 or early 1867, a 15-year-old boy named Erasmus Jacobs picked up a large yellowish pebble near the source of the Orange River in South Africa. He figured his sister would enjoy playing with it. A month later, when Erasmus and his sister were playing a game called five stones with it and other river stones, a farmer named Schalk van Niekerk arrived. He saw the shiny stone and thought it might be of some value, so he asked their mother to sell it to him. She laughed at the idea of selling a stone and told him he could have it for nothing. Van Niekerk believed it could be a diamond thanks to a book on precious stones he had received and a suggestion that the area might contain diamonds. He asked a trader friend named John O'Reilly to find out if it was actually a diamond.

O'Reilly showed the stone to many people in Hopetown and said he thought it was a diamond, but they all laughed at him. A few in the town thought it was topaz because of its yellowish color and, therefore, of little value. Then he showed it around Colesberg. Most ridiculed the idea that the stone was a diamond, except the acting civil commissioner, Lorenzo Boyes, who had it tested by an expert in mineralogy. The mineralogist confirmed it was a diamond and estimated its value to be £500.

Later the stone was sent to England to be tested by the crown jeweler with the intention of sending it on to be displayed at the Cape Colony's stand at the 1867 Paris Exhibition, which was going to be opened by the Emperor Napoleon III on April 1. After the crown jeweler declared it a diamond, the stone or possibly a glass replica was exhibited in Paris, but doubts were expressed about whether the diamond could really be from South Africa, especially after a search near where it was found failed to turn up any other diamonds, although more were eventually found a couple of years later.

The stone that was at first considered just a plaything is now recognized as the first authenticated diamond found in South Africa. It was named the Eureka, which means "I have found it" in Greek. It originally weighed 21.25 carats but was later cut into a 10.73-carat cushion-shaped brilliant. For the 100th anniversary of its discovery, the mining company De Beers purchased the Eureka and donated it to the South African people. The diamond was placed in the Kimberley Mine Museum, where it is now on display.

## A Mineral Composed of Carbon with a Unique Crystal Structure

To a chemist, a diamond is a mineral consisting of carbon atoms arranged in a cubic crystalline structure. Other substances, such as graphite, have the same chemical composition, but their carbon atoms are arranged differently. Graphite's atoms are arranged in layers. Its atomic bonds within the layers are fairly strong, but the bonds between the layers are weak. In contrast, diamond's atomic bonds are strong in all directions. This is why diamond is so hard.

Diamond's crystallization process begins with a carbon atom, which bonds with other carbon atoms. The bonded atoms form groups of five, called tetrahedrons, which have one carbon atom in the center and the other four atoms surrounding it. Tetrahedrons, in turn, combine to form a unit cell, and many unit cells together form a diamond crystal.

# The Chemical Composition and Structure of Diamond

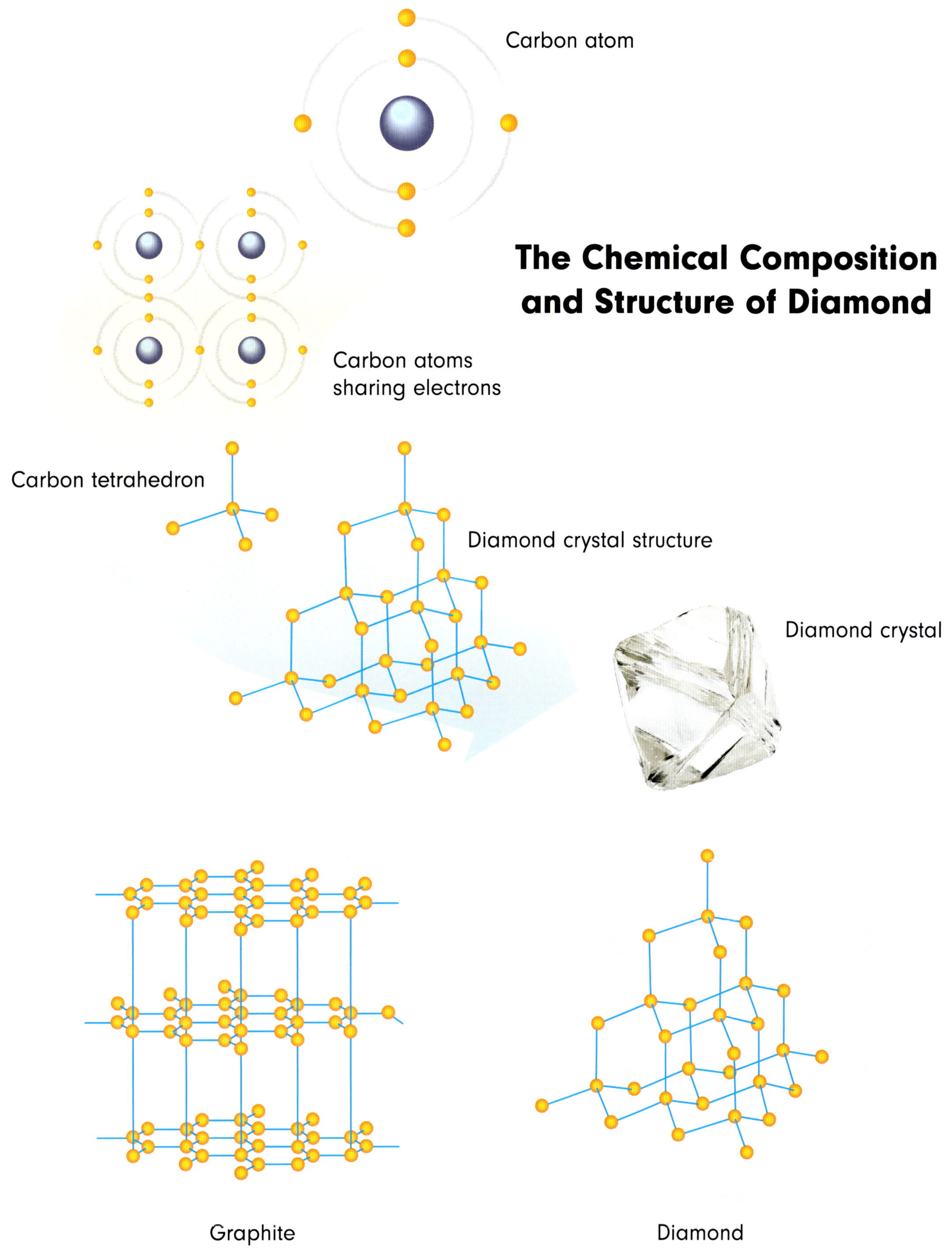


*Illustrations adapted from GIA images by Peter Johnston*

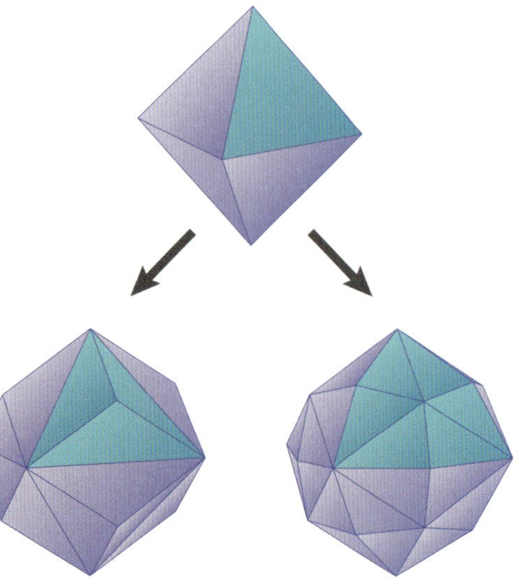
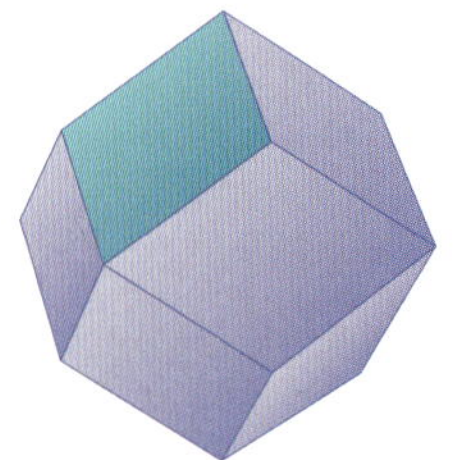

Diamond crystal shapes: Many diamond crystals are actually modifications of the cube or octahedron. The blue areas in the images show where the faces of simple and complex crystals correspond. *Image by Peter Johnston © GIA; reprinted with permission*

A transparent octahedral diamond crystal. *Diamond courtesy of Pala International; photo by Mia Dixon*

A square diamond crystal. *Diamond and photo courtesy of The Arkenstone; photo by Joe Budd*

A diamond's unit cell is cube shaped, and it can be the basis of different shapes. The basic crystal shape of a mineral is called its habit, and the habit of a gem-quality diamond is often the octahedron, a shape with eight triangular sides. However, perfectly shaped octahedral rough is rare, so it often ends up as a collector's specimen rather than a cut gem. A dodecahedron has 12 square flat surfaces, but most dodecahedral diamond crystals are rounded and are well suited for fashioning round brilliants.

Crystal structure affects diamond in ways that are important to cutters. As you will see in chapter 4, the unique chemical structure of each diamond crystal determines how and into what shape it is cut.

Though all diamonds are composed of carbon, other atoms, called impurities, can be found in a diamond's chemical structure. For example, nitrogen, which is the most common impurity, causes yellow coloration. Other factors, like irregularities and defects in the chemical structure or radiation, can affect diamond color as well, but nitrogen is the most frequent source of diamond color.

In 1934 a study by R. Robertson, J.J. Fox and A.E. Martin showed that diamond falls broadly into two categories they called Type I and Type II, based on their transparency to ultraviolet and infrared wavelengths. Further research between 1954 and 1959 showed that the differences between these two categories were due to the presence of nitrogen in the diamond structure of Type I diamonds and the apparent absence of nitrogen in Type II diamonds.

Diamond crystals are found in a variety of shapes and colors. *Diamonds courtesy of Pala International; photo by Mia Dixon*

Rough diamond from the Lulo diamond mine in Angola. The large crystal on the left is a macle – a flat, triangular twinned diamond crystal. *Photo © Lucapa Diamond Company Limited*

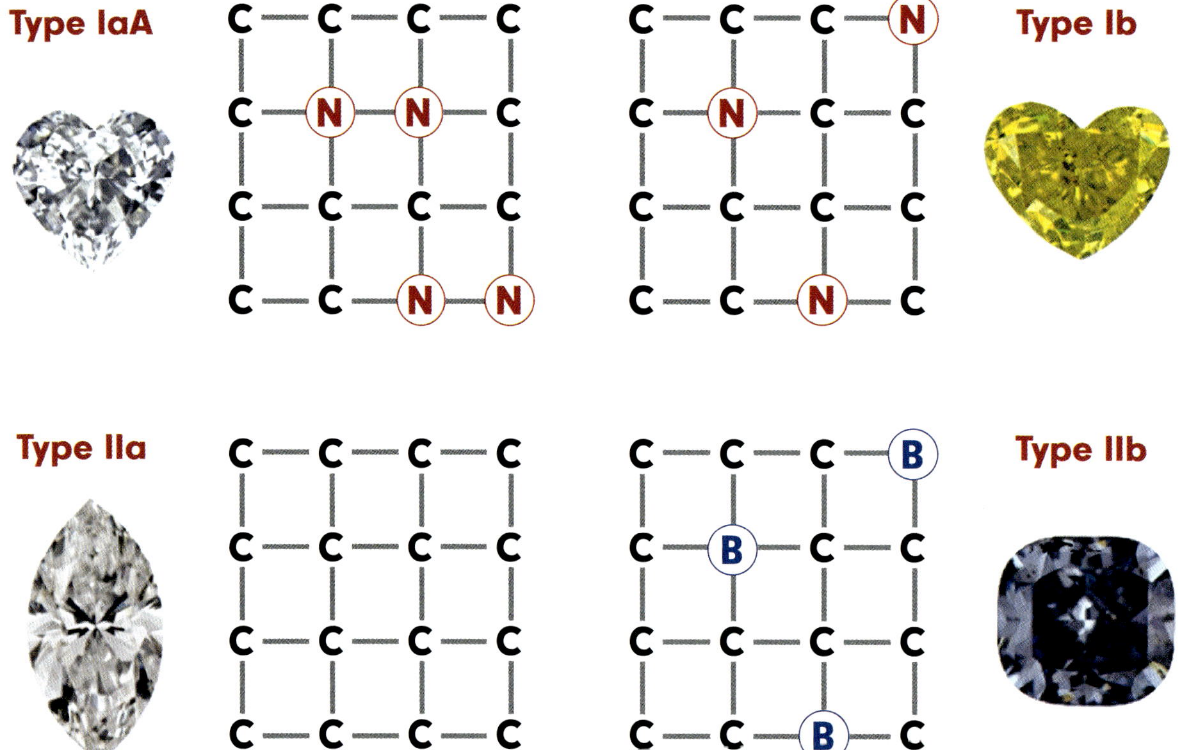


This diagram shows a simplified version of the diamond type classification system. Type I (top row) and Type II (bottom row) diamonds can each be divided into two subcategories based on the arrangement of carbon and impurity atoms in the diamond structure. C = carbon atom, N = nitrogen atom, and B = boron atom. Diamond type can be determined quickly with a scientific method called infrared spectroscopy. *Diagram by James Shigley and Mike Breeding © GIA; reprinted with permission*

Type I diamonds are further divided into two types: Type Ia has nitrogen atoms in pairs or clusters, and Type Ib has fewer nitrogen atoms than Type Ia, and the nitrogen atoms are scattered and isolated. The 2003 GIA Diamonds & Diamond Grading course estimates that about 95 percent of cuttable-size natural diamonds are near-colorless to yellow Type Ia diamonds. Brown, orange, pink and green natural diamonds may also be Type Ia. Lab-grown diamonds, however, are not Type Ia when grown.

Type Ib diamonds generally have a stronger yellow color than most Type Ia diamonds, but Type Ia diamonds may also have a vivid yellow color. Type Ib diamonds can also be orange or brown.

Type II diamonds contain very little or no nitrogen, and they are rare. Most Type IIa diamonds are colorless, but crystal distortion can make them brown or gray. In addition, some green and pink diamonds are Type IIa. The majority of the world's most famous colorless diamonds are Type IIa, including the Koh-i-noor and the Cullinan. Type IIa diamonds are excellent conductors of heat.

Type IIb diamonds contain boron, which turns them blue and makes them excellent conductors of electricity. The Hope Diamond is Type IIb. Gray diamonds can also be Type IIb.

| Basic color | Causes of color |
|---|---|
| ◆ Red | Irregularities in the crystal's atomic structure as a result of deformation |
| ◆ Violet | Hydrogen impurities |
| ◆ Purple | Irregularities in the crystal's atomic structure as a result of deformation |
| ◆ Green | Natural radiation, hydrogen; green fluorescence may occasionally make a diamond appear greenish |
| ◆ Blue | Boron impurities, sometimes radiation |
| ◆ Orange | Probably chemical impurities and/or structural distortion |
| ◆ Pink | Structural defects combined with various impurities of nitrogen or hydrogen |
| ◆ Yellow | Isolated nitrogen atoms that randomly take the place of individual carbon atoms or aggregates (clusters) of three nitrogen atoms |
| ◆ Green Yellow | Natural radiation, hydrogen, nitrogen |
| ◆ Olive* | Natural radiation, hydrogen, nitrogen |
| ◆ Black | Black inclusions |
| ◆ Brown | A defect in the atomic structure of the crystal probably caused by tremendous pressure and indicated by colored graining |
| ◆ Gray | Hydrogen impurities |
| ◆ Chameleon** | Natural radiation, hydrogen and nickel impurities; cause of color change uncertain, though nitrogen, nickel and/or hydrogen may be involved |

Sources: *Gems & Gemology in Review: Colored Diamonds*, edited by John M. King; "Nature of Color in Diamonds" by Emmanuel Fritsch in *The Nature of Diamonds*, edited by George Harlow; *Collecting and Classifying Coloured Diamonds* by Stephen C. Hofer; and NCDIA.com.
* Olive: a common color term for grayish-yellowish green or grayish-greenish yellow used by some dealers but not GIA.
** Chameleon diamonds show a color change typically from olive green to brownish yellow when heated to 302°F (150°C) or when stored in darkness for a few days, but the reverse change is possible.

Nitrogen and boron impurities are not the only causes of color in diamonds. Hydrogen impurities can make diamonds violet, gray or grayish. Natural radiation can produce green or greenish colored diamonds, and fluorescence can contribute to diamond color. Inclusions can make diamonds look black or white. The table above briefly identifies the causes of color in various colored natural diamonds. Keep in mind that the process of creating color in diamonds is a lot more complex than the table makes it appear. Examples of diamonds with some of these colors are on page 30.

Fancy Vivid bluish green

Fancy Intense yellow

Fancy Intense green

Fancy Intense greenish yellow

Fancy pink

Even though many diamonds have single colors like yellow, pink and green, diamonds often have modifying colors like bluish and greenish. Keep in mind that the printing and developing processes usually alter the true color of gems in photographs. *Diamonds and photos courtesy of Joe Namdar of Namdar Diamonds*

Diamonds in the normal color range. Do not use this photo to grade the color of diamonds. The printing and developing processes and paper color usually alter the true color of gems in photographs. *Photo by Tino Hammid © GIA; reprinted with permission*

Most diamonds range in color from colorless to light yellow, gray or brown. This is known as the normal color range. It is also called the D-to-Z range because of the letter designations given to each step in the color scale of the Gemological Institute of America (GIA), the most widely used diamond color scale. Diamonds ranging in color from E to O are shown in the photo above.

Diamonds beyond the normal color range are called fancy color diamonds and can be various shades of yellow, brown, orange, red, pink, violet, red, blue or green. GIA introduced the term fancy to describe naturally colored, faceted diamonds that exhibit either noticeable yellow or brown or any color other than yellow or brown. See chapter 6 for additional information on diamond color and how it affects the price of diamonds.

## A Direct Sample of the Interior of the Earth

To a geologist, a diamond is a glimpse into the interior of the Earth. Many diamonds contain mineral particles and crystals (known as inclusions) that allow scientists to determine the composition of the Earth and its geological processes at depths of more than 62 miles (100 kilometers) underground and often from billions of years in the past (*Gems & Gemology*, Winter 2013). No other mineral sample can provide information for research from such depths and from so long ago. Diamond is an artifact of our planet.

Most natural diamonds form under ancient, deep and stable parts of the Earth's crust called cratons, which make up the interior portions of continents and extend down at least 93 to 124 miles (150 to 200 kilometers). These billion-year-old cratons all have thick continental roots with cool temperature profiles conducive to diamond formation. Examples of cratons are the Siberian craton in Russia, the Slave craton in Canada and the Kalahari craton in South Africa and Botswana. Diamonds that form within these continental roots are called lithospheric diamonds. During rare volcanic eruptions, these diamonds are carried up to the surface of the Earth in an igneous rock called kimberlite.

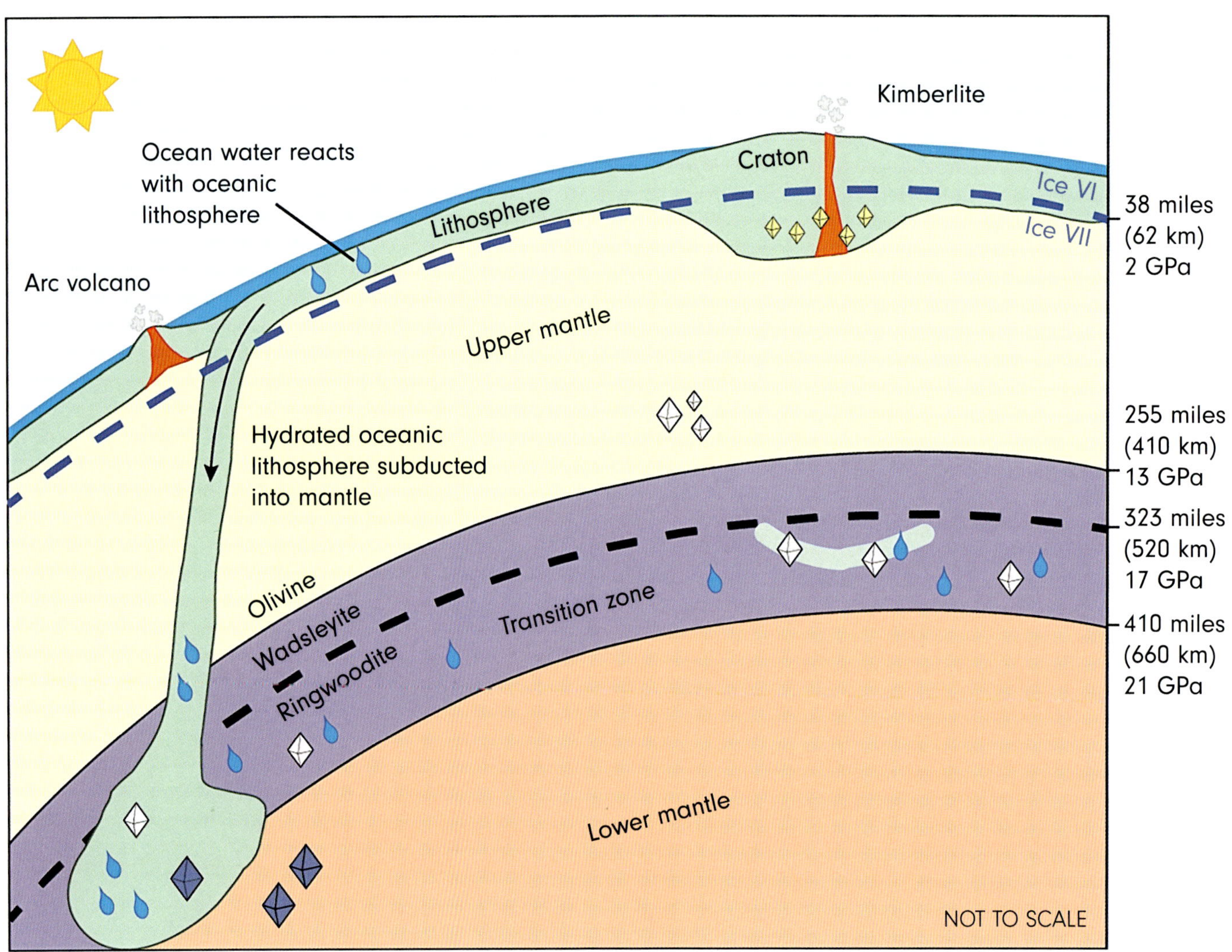


The majority of diamonds come from the cratonic lithosphere (shown here as yellow diamonds). Superdeep diamonds (shown here as white and blue diamonds) are rarer and originate from greater depths, often from the transition zone. *From the Summer 2018 issue of* Gems & Gemology*; diagram © GIA; reprinted with permission*

Diamonds called superdeep diamonds form in the areas of the mantle known as the transition zone, which is 255 to 410 miles (410 to 660 kilometers) below the Earth's surface, and the lower mantle, which sits below the transition zone. After formation, the diamonds are transported to shallower depths of the mantle, likely through mantle convection cells. They can also be brought to the surface by kimberlite eruptions (*Gems & Gemology*, Summer 2018). Many of the world's most exceptional diamonds, such as the 3,105.75-carat Cullinan diamond, are superdeep diamonds. They tend to be relatively pure and inclusion poor and have an irregular shape in rough form (*Gems & Gemology*, Winter 2017).

## A Source of Income

For artisanal miners in countries like Sierra Leone, where diamonds are found in rivers and streambeds, diamonds are a source of income and a way out of poverty.

People in Botswana have benefited indirectly from the diamonds discovered there in the late 1960s. Diamond profits have been used to build schools and hospitals. At the time of its independence from the United Kingdom, in 1966, Botswana was among the poorest countries in the world. Thanks to diamonds mined there, it has become a middle-income country. Diamonds now represent about 60 to 80 percent of its export earnings. According to Festus Mogae, the third president of Botswana, "every diamond purchase represents food on the table, better living conditions, better healthcare, potable and safe drinking water and more roads to connect remote communities."

## An Insatiable Obsession

Harry Winston (1896–1978), a famous diamond wholesaler known as the King of Diamonds, described what diamonds were to him in his introduction to *The Book of Diamonds* (1965) by Joan Younger Dickinson:

> *Today jewels are more than my love and my life; they are an insatiable obsession. It is not only the lure of hidden beauty that lies in a rough diamond; there is also the excitement of speculating — and the reward of having my judgment proved right.*
>
> *Recently I was offered a 39-carat diamond in the rough by 10 men who had pooled their money to invest in it. It was not a bargain by any means, and it had a flaw right down the middle. But when I looked at it, I realized that with proper cutting, that flaw would come completely out; I could see exactly what could be done and should be done to create a great pear-shaped gem.*
>
> *I bought it. For weeks the cutting obsessed me, but when it was finished, we had an exquisite stone, perfectly cut, and I felt the pride of an adoring father.*

A portrait of legendary jeweler Harry Winston. *Photo by Alfred Eisenstaedt/The LIFE Picture Collection via Getty Images*

A portrait of "Diamond" Jim Brady.
*Everett Collection Historical/Alamy Stock Photo*

Diamonds also obsessed Jim Brady (1856–1917), a self-made millionaire who was born above the saloon his father owned in the west side of Manhattan. Even though he had humble beginnings, Brady became a very successful businessman known for his love of diamonds and jewelry. He had diamonds on his buttons, watch, belt buckle, scarf pin, eyeglass case, rings, tie pins, walking cane, cuff links and even on the buttons of his underwear, so people called him Diamond Jim. In 1895 he also became known for being the first person in New York City to own an automobile. Diamond Jim believed that a man of wealth should show off his prosperity and be generous. Nurses who cared for him in hospital got diamond rings. After living a full life, he died in his sleep at the age of 60. Most of his fortune was left to the New York Presbyterian Hospital and to Johns Hopkins Hospital in Baltimore, Maryland.

## A Form of Portable Wealth

Diamonds are the most concentrated and portable form of wealth in existence. At times, they have been used by refugees to escape totalitarian regimes. For example, leading up to and during World War II, many Jews who escaped Nazi Germany used diamonds to pay their way out of the country and start a life elsewhere.

Unlike diamonds, real estate cannot be transported. Gold, though portable, is not as valuable when compared to diamonds in terms of wealth concentration. Gold sold for as much as $2,067 an ounce in 2020, but a few high-quality diamonds totaling the same weight can be worth millions of dollars.

Some wealthy people today use high-quality diamonds as an additional means of diversifying their financial portfolios, especially in countries where governments have devalued currencies or replaced their banknotes, removing the old notes from circulation. However, many of these investors do not necessarily store their diamonds in safety deposit boxes. They may have them set in jewelry so that they can get continual enjoyment from them, which is arguably a better reason to buy diamonds.

## A Symbol of Love and Commitment

Archeological evidence suggests that about 4,800 years ago Egyptians exchanged rings of love, often made of materials such as twisted reeds or hair. The circle was a symbol of eternity and never-ending love, and the hole in the center was a gateway leading to events known and unknown.

The first written evidence of engagement rings dates back to the Romans around 200 BCE. Their rings, however, were more of a business contract and sign of ownership than a symbol of love. Roman men would claim their women by giving them a ring. *Fede* rings depicting two hands clasped together became popular. Metals, such as bronze, copper, silver, iron and gold, were used to make these rings.

It was not until 850 CE that the engagement ring was given an official meaning. Pope Nicholas I declared that the engagement ring signified a man's intent to marry. Gold was the most common material for engagement rings at this time.

The earliest documented use of a diamond set in an engagement ring is 1477, when Archduke Maximilian I of Austria ordered a diamond ring for his bride-to-be, Princess Mary of Burgundy. Because of the royals' public standing, diamond engagement rings became a custom among other wealthy couples, too, in the same way that celebrities inspire trends today. Mary's ring signified her everlasting faithful devotion to her husband, but she sadly died after just five years of marriage, from a horse-riding accident.

Diamonds of the 1400s were not bright and sparkly like those of today. Early cutting techniques caused gems to look dull and even black. Compensating for these lackluster stones, goldsmiths designed elaborate settings, composed of romantic themes, such as rosettes and fleur-de-lis. Mary of Burgundy had an ornate gold ring that was set with irregular baguette-like dark diamonds in the form of her initial, M.

The unusual hardness of diamond, which kept it from being deformed or abraded, eventually made it a symbol of unwavering love that would endure forever. To the upper-class Europeans of the Renaissance, diamonds represented fortitude, faithfulness, innocence and prosperity. They assumed that a diamond's power to withstand natural forces could be transferred to the owners and that they, in turn, would be able to withstand temptation.

The ring of Mary of Burgundy. *KHM-Museumsverband*

A colorized engraving of Mary of Burgundy and Maximilian I. *PRISMA ARCHIVO/Alamy Stock Photo*

# a girl's dream

**_TREASURED IN A DIAMOND_**

A girl, hopefully, tenderly, dreams upon a fresh, new miracle, her 'wakened love. Her heart brims with thoughts of "him," alone. And their engagement diamond, telling their promise, records this dear detail. So, though the enchantment of the engagement time comes but once, it will never be forgotten. The earth-born firelight he's placed upon her finger will recall their love's first meaning, endlessly. To mark your engagement, your ring-stone may be modest in size, but it should be chosen with care, for it will be treasured always, by you and generations yet to be.

**_HOW TO BUY A DIAMOND_** First, and most important, consult a trusted jeweler. Ask about color, clarity and cutting—for these determine a diamond's quality, contribute to its beauty and value. Choose a fine stone, and you'll always be proud of it, no matter what its size. Diamond sizes are measured by weight, in points and carats—100 points to the carat. (Exact weights shown are seldom found.) For your guidance, the price ranges given below are based on current quotations by jewelers throughout the country in July, 1961. Note that diamond prices vary widely according to the qualities being offered. Tax additional.

Lovely dreamer . . . painted for the De Beers Collection by Edwin Schmidt

De Beers Consolidated Mines, Ltd.

25 points (¼ carat)
$75 to $225

50 points (½ carat)
$150 to $590

1 carat (100 points)
$400 to $1680

2 carats (200 points)
$840 to $4000

a diamond is forever

It was not until the discovery of the large deposits in South Africa that diamonds began to reach a market beyond royalty and the wealthy elite. By the end of the 19th century, the middle class were beginnning to wear diamonds. As the output of the African mines increased, the mining company De Beers realized advertising would be needed to promote diamonds, of which it controlled the world's supply. In 1938 Harry Oppenheimer, the chairman of De Beers Consolidated Mines, met with the president of N.W. Ayer & Son, a Philadelphia-based advertising agency, to develop a strategy for promoting diamonds. Ayer suggested that instead of promoting an elitist image of diamonds, it would be better to focus on a broader market segment: lovers. Thus began one of the most successful and enduring partnerships in advertising history.

Ayer first used Hollywood stars for the campaign. They were shown wearing diamonds in movies and advertisements. Ayer also relied on jewelers, lecturers and magazines to encourage the public to wear diamonds. However, a slogan was needed. In 1948 a copywriter at Ayer by the name of Frances Gerety was working late one night. She put her head down on the table to rest, but before she left work she scribbled the words "A Diamond Is Forever" on a piece of paper. That simple yet effective phrase became America's most famous advertising slogan.

Another phrase that made history is "Diamonds Are a Girl's Best Friend," which was introduced as a jazz song by Carol Channing in the 1949 stage musical *Gentlemen Prefer Blondes*. A movie comedy based on the play was released in 1953 starring Marilyn Monroe. Her rendition of the song "Diamonds Are a Girl's Best Friend" was a stellar performance and was ranked as the 12th best film song of the 20th century by the American Film Institute.

N.W. Ayer's advertising campaign played a major role in promoting diamonds for engagement rings, but there are practical reasons why diamonds are an ideal choice for rings. Diamond is more resistant to scratching, abrasion, heat and chemicals than any other gem, and its neutral color allows it to be worn with any color of clothing. Diamonds are also readily available. As a result, diamond remains the most popular gem for engagement rings and continues to signify commitment and enduring love.

A diamond-clad Marilyn Monroe in the 1950s. Monroe's iconic role in *Gentlemen Prefer Blondes* cemented her status as a rising Hollywood star. *Everett Collection Inc/Alamy Stock Photo*

(opposite page) A 1961 ad from De Beers, featuring the iconic slogan "A Diamond Is Forever." The image was painted by Edwin Schmidt. *Neil Baylis/ Alamy Stock Photo*

# 2

# Where Are Diamonds Found?

Erasmus Jacobs, the 15-year old son of a poor farmer named Daniel Jacobs, is credited with the first authenticated diamond discovery in Africa. He found it on a farm near the Orange River in South Africa sometime between December 1866 and February 1867. The stone was eventually sent to a mineralogist, who verified it to be a diamond weighing about 21.25 carats and worth £500 in rough form. Later the stone made its way to London and was cut into a 10.73-carat diamond named the Eureka. See chapter 1 for more information.

After the discovery of the Eureka was reported, a few diamond hunters found their way to the area around the Orange and Vaal Rivers in central South Africa, and a geologist named James A. Gregory did a lengthy study of the area. In December 1868 he reported in London's *Geological Magazine* that he saw no indications that would warrant expectation of finding diamonds or diamond-bearing deposits there. He concluded that "the whole diamond discovery in South Africa is an imposture — a bubble scheme." This conclusion soon turned out to be false.

In March 1869 a shepherd picked up an interesting-looking pebble. He tried to barter the stone, and everyone turned him down except Schalk van Niekerk, the local farmer who had been involved in the discovery of the Eureka. Van Niekerk gave the shepherd a horse, 10 oxen and 500 sheep in exchange for the stone, which was four times the size of the Eureka. He later sold it for £11,200.

The 83.50-carat diamond was named the Star of South Africa and was displayed in the South African Parliament Building. The gem was fashioned into a 47.69-carat pear-shaped brilliant and was sold to the Earl of Dudley.

The Star of South Africa started a diamond rush that brought people from Europe, North America and Australia to South Africa in search of riches. Within seven years of the gem's discovery, South Africa was producing more than 90 percent of the world's diamonds. Today less than 10 percent of natural diamonds come from South Africa, with countries like Russia, Botswana and Canada each producing more.

This chapter will explore where diamonds are mined, from the earliest mines in India to today's large-scale operations in resource-rich Russia.

## Basic Diamond Mine Terminology

**Rough:** A natural, uncut diamond.

**Primary deposit:** A deposit where rough is embedded in hard rock, which is called host rock.

**Secondary deposit:** A deposit where rough is found away from its primary source.

**Alluvial deposit:** A secondary deposit where rough has accumulated in the sand and gravel of dried-up riverbeds, rivers, streams and ocean shores as the result of erosion and weathering of host rock.

**Alluvial:** Any diamond that is found in an alluvial deposit.

**Marine deposit:** A secondary deposit where rough is found with soil and rock particles that have been transported from land areas to the ocean floor or shoreline by wind, ice or rivers.

**Igneous rock:** Rock that starts out in a molten or partially molten state and becomes solid when it cools.

**Kimberlite:** An igneous rock that transports diamonds to the surface of the Earth. Kimberlite is the host rock for most primary-deposit diamonds.

**Lamproite:** An igneous rock that transports some diamonds to the surface of the Earth. Lamproite is the source rock for diamonds in Western Australia and Arkansas.

**Diamond pipe:** A deep, carrot-shaped formation resulting from a violent eruption that blasts kimberlite and lamproite (containing diamonds) up through the Earth's crust, forming a crater at the surface with a long, vertical pipe below it.

## INDIA

India was the first country to supply the world with diamonds and the only significant diamond source until about 1730, when Brazil replaced India as the world's main source. Diamonds were found in Indian riverbeds some time before 600 BCE, and Alexander the Great may have brought the first diamonds to Europe in 326 BCE, after he invaded northern India.

The most famous diamond mining area in India is the Golconda region in the present-day states of Telangana and Andhra Pradesh. The alluvial mines in this area, particularly the Kollur mine, produced some of the largest and finest diamonds in the world, including the Koh-i-noor, the Dresden Green, the Orlov, the Regent and the Hope Diamond. As a result, "Golconda diamond" has become an industry term that refers to exceptional diamonds with unusually high transparency. Diamonds from the Golconda mines were taken to the city of Hyderabad to be sold. Some of the most prized diamonds, such as the Koh-i-noor, were kept under guard in the city at the Golconda Fort, which is now a monument of national importance in India.

One of the best sources of information about the Indian mines is *Les Six Voyages de Jean-Baptiste Tavernier*, which was published in 1676. It is a compilation of Jean-Baptiste Tavernier's experiences during his six trips to India as a French gem merchant. Tavernier is best known for buying the 115-carat Tavernier Blue diamond, which he sold to Louis XIV of France in 1668. Louis had the court jeweler recut the stone into the 69-carat French Blue, but during the French Revolution it was stolen and then recut again. It was eventually purchased by collector Henry Philip Hope and renamed the Hope Diamond.

Diamond production in India probably reached its peak in the 1600s. Tavernier reported that there were 60,000 men, women and children at work at the diggings in one area. By about 1750 Indian production was insignificant, though an occasional large stone was found there. The Panna region in the state of Madhya Pradesh is the only mining area in India producing diamonds commercially today.

Since the 1970s India has been the world's major cutting center for small diamonds, and it has become an important cutting center for larger diamonds as well. It is estimated that 14 out of 15 diamonds in

The approximately 41-carat Dresden Green diamond is mounted in a hat ornament and on display at the Green Vault in Dresden Castle, Dresden, Germany. *Wikimedia Commons*

PAKISTAN
HARYANA
DELHI
New Delhi
NEPAL
RAJASTHAN
UTTAR PRADESH
BIHAR
Panna
JHARKHAND
GUJARAT
MADHYA PRADESH
INDIA
Surat
CHHATTISGARH
ODISHA
DAMAN & DIU
DADRA &
NAGAR HAVELI
Mumbai
MAHARASHTRA
TELANGANA
Hyderabad &
Golconda Fort
Kollur
GOA
ANDHRA
PRADESH
ARABIAN SEA
BAY OF BENGAL
KARNATAKA
PUDUCHERRY
LAKSHADWEEP
TAMIL NADU
KERALA

A depiction of an Indian diamond mine published in the first volume of *La Galerie Agréable du Monde* (the pleasant gallery of the world), published by Pieter van der Aa in Leiden, Netherlands, in 1725. *Wikimedia Commons*

the world are cut and polished in India. India's cutting industry is centered in the states of Gujarat and Maharashtra, near the international trading center of Mumbai (formerly Bombay). The city of Surat, in the state of Gujarat, is famous for its diamond cutting and polishing and is known as the Diamond City of India. About 90 percent of the world's diamonds are polished there. Another diamond bourse (exchange) is scheduled to be operational in Surat by 2022.

India is also an important manufacturing center for diamond jewelry. As a result, there are more than one million people employed by the diamond industry in India. The lives of people around the diamond centers have improved significantly thanks to the medical facilities, schools and employment that the diamond industry has contributed.

Diamond sorting at the Dharmanandan Diamonds cutting facility in Surat, India. *Photo © Dharmanandan Diamonds Pvt. Ltd.*

Workers hand polish diamonds. *Photo © Dharmanandan Diamonds Pvt. Ltd.*

The Banjarmasin Diamond, which is currently on display at the Rijksmuseum in Amsterdam, Netherlands. *Everett Collection Inc/Alamy Stock Photo*

Traditional diamond mining in Borneo. *agefotostock/Alamy Stock Photo*

## Borneo (Kalimantan, Indonesia)

Borneo is an island that is politically divided among three countries: Malaysia and Brunei in the north and Indonesia to the south. Diamond mining in Borneo may have started as early as 600 CE.

The first written reference to diamond mining on the island was by Duarte Barbosa of Portugal in 1518. The Dutch colonized Borneo in the early 1600s and began exploiting diamonds through the Dutch East India Company (*Gems & Gemology*, Summer 1988). The diamonds acquired by the Dutch East India Company were exported to the Netherlands, and they helped lay the foundation for the development of Amsterdam as an international diamond cutting and trading center. The discovery of the diamond deposits of South Africa led to the decline of the Borneo diamond fields, though some minimal mining is still done there today.

Most of Borneo's diamonds have come from the large Indonesian portion of the island, which is called Kalimantan. One of the more notable diamonds is the Banjarmasin, which was once owned by the sultan of Banjarmasin, now the capital of South Kalimantan. Originally the diamond was an octahedron with an estimated weight of 70 to 77 carats. In 1859 the Dutch seized control of the city and looted its treasures, including the Banjarmasin Diamond. The rough diamond was sent to the Netherlands and cut into a 38.25-carat diamond resembling an old mine cut. As of the writing of this book, the Banjarmasin Diamond is on display at the Rijksmuseum in Amsterdam, but it may eventually be returned to Indonesia.

The Borneo diamond deposits are alluvial and found primarily in western Kalimantan along the Landak River, near the equator, and in southeastern Kalimantan under the Danau Seran Swamp. Danau Seran is near Martapura, the largest diamond cutting center in Indonesia, which is about 24 miles (39 kilometers) southeast of Banjarmasin, the capital of South Kalimantan. Tours of the Martapura diamond market and nearby Cempaka mining area are available. Much of the mining is done using traditional methods, such as panning. Panning for diamonds takes place at the same time as panning for gold and other gems.

The majority of Borneo's diamonds are gem grade. According to colored diamond dealer Arthur Langerman's website, "Production on the island mainly consists of diamonds tinted with yellow and brown, but the red, blue and green diamonds that can also be found there are among the most exceptional in the world."

SOUTH CHINA SEA
BRUNEI
MALAYSIA
CELEBES SEA
NORTH KALIMANTAN
BORNEO
Landak River
WEST KALIMANTAN
INDONESIA
EAST KALIMANTAN
MAKASSAR STRAIT
CENTRAL KALIMANTAN
SOUTH KALIMANTAN
Banjarmasin
Martapura
Danau Seran & Cempaka
JAVA SEA

The Moussaieff Red. *Diamond and photo: Moussaieff Jewellers*

Carbonado diamond suspended from platinum hoops set with white and black diamonds. *Earrings and photo by Jim Grahl*

## Brazil

Diamonds were discovered in Brazil in the early 1700s by miners looking for gold along the Jequitinhonha River in the state of Minas Gerais, near the village of Arraial do Tijuco (later named Diamantina). The official discovery of diamonds by the Portuguese, who occupied Brazil, is recorded as 1725. By 1730 Brazil replaced India as the world's main source of diamonds and kept that position until 1870, when production started in South Africa. The majority of the diamonds in European jewelry from the mid-18th century to the mid-19th century originated from Brazil.

Between 1730 and 1735 the Brazilian diamond market grew so much that the price of diamond rough dropped dramatically. Curiously, while the price of rough fluctuated enormously with variations in supply, the price of cut stones remained stable.

Most of Brazil's diamonds have been found as loose crystals in rivers or streams or in sediments nearby. These alluvial deposits occur throughout Brazil, but those in the states of Minas Gerais and Mato Grosso have been the most economically important over the past three centuries (*Gems & Gemology*, Spring 2017).

Most Brazilian diamonds are recovered by independent miners called *garimpeiros* using simple tools and pans. The world's largest known Fancy red diamond, the Moussaieff Red, was found as a 13.90-carat crystal by a Brazilian farmer in an alluvial deposit in 1989 and cut into a 5.11-carat triangular brilliant. Many other large fancy color diamonds have also been found in Brazil.

Brazil is also noted for its carbonado, a polycrystalline diamond composed of many tiny black crystals of diamond and graphite that are randomly oriented with respect to each other. Carbonado diamond is black, opaque, porous, tougher than single crystal diamond and extremely difficult to cut and polish, so most of it is sold as rough and for industrial use. However, some of it is faceted. Brazil and the Central African Republic are the only known sources of carbonado.

In 1895 a 3,167-carat carbonado diamond was found above ground in Lençóis, in the Brazilian state of Bahia, by Sérgio Borges de Carvalho. Named the Sergio, it is believed to have come from outer space and is the largest carbonado diamond ever found. The

An illustration depicting 19th-century diamond mining in Brazil. The engraving is from *Underground Life; or Mines and Miners* (1868) by L. Simonin. *Sheila Terry/ Science Photo Library*

A ballas diamond from Minas Gerais, Brazil. *Diamond and photo courtesy of The Arkenstone; photo by Joe Budd*

Sergio was first sold for $16,000 and later for £6,400 and was broken up into 3- to 6-carat pieces for use in industrial diamond drills.

Ballas is another type of polycrystalline diamond found in Brazil. Like carbonado, it is an industrial diamond that is tougher than single crystal diamond and very hard to cut and polish, but it is found in a wider range of colors. Most are gray; others can be yellowish, brown or white.

Kimberlite pipes have been found in Brazil. In July 2016, after a few years of exploration and development, Lipari Mineraçao Ltda started commercial production at a kimberlite field called the Braúna complex. It is in the northeastern state of Bahia and is South America's largest diamond mine.

Brazil produces less than 1 percent of the world's diamonds today. However, it has yielded a number of large diamonds as well as fancy color and superdeep diamonds and carbonado. With large-scale kimberlite mining in the Bahia state, its production could grow (*Gems & Gemology*, Summer 2017).

## South Africa

In 1869 thousands of diamond prospectors from around the world went to central South Africa after the Star of South Africa diamond was discovered there. They set up claims on the banks of the Vaal River near where the diamond had been found and dug the riverbed area for diamonds, using mostly picks and shovels.

In late 1869 diamonds were discovered about 12 miles (20 kilometers) from the nearest Vaal River deposits, far from any rivers or streams. This in turn led to the discovery of kimberlite pipes and the realization that diamonds were not just found in alluvial gravel, as they had been in India, Borneo and Brazil. Several important pipes were clustered together — Dutoitspan, Bultfontein, De Beers and Kimberley. Kimberley, the area in which these kimberlite pipes were found, became the capital of the world diamond industry. Today it is a city of more than 200,000 people.

An overhead view of a diamond mine in Kimberley, South Africa, dating from the late 1800s. *Chris Howes/Wild Places Photography/Alamy Stock Photo*

Cecil Rhodes founded De Beers in 1888.
*Lebrecht Music & Arts/Alamy Stock Photo*

In 1874 there were hundreds of claims in the Kimberley area. Little by little, claim holders from South Africa, Canada, Australia and Europe joined forces and formed partnerships and companies to finance powerful machinery, which gave rise to a boom in South African diamond production. This resulted in a serious crisis of oversupply between 1873 and 1894, which sent diamond prices tumbling and lowered the value of the claims.

Cecil Rhodes (1853–1902), an Englishman and claim holder at the De Beers mine, took advantage of the crisis and, together with his partners, bought up claims at low prices. By 1887 Rhodes had gained complete control of the De Beers mine. Eventually Rhodes and his associates were able to buy up other claims and have majority ownership of the mines in Kimberley. In 1888 Rhodes incorporated his new company and named it De Beers Consolidated Mines Limited. By 1900 De Beers controlled about 90 percent of the world's production of rough diamonds.

Shortly after De Beers was formed, it signed a contract with a group of prominent London diamond merchants known as the London Diamond Syndicate. It contracted to buy and sell all of the output of

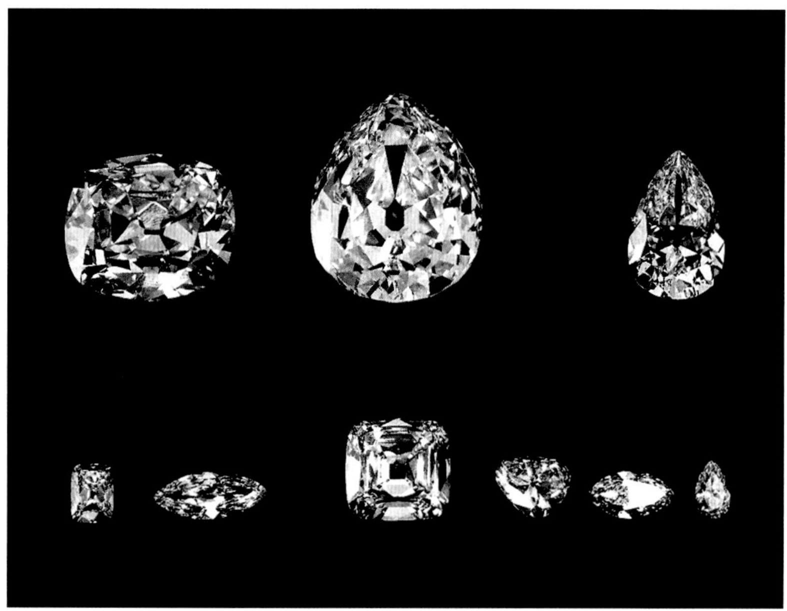

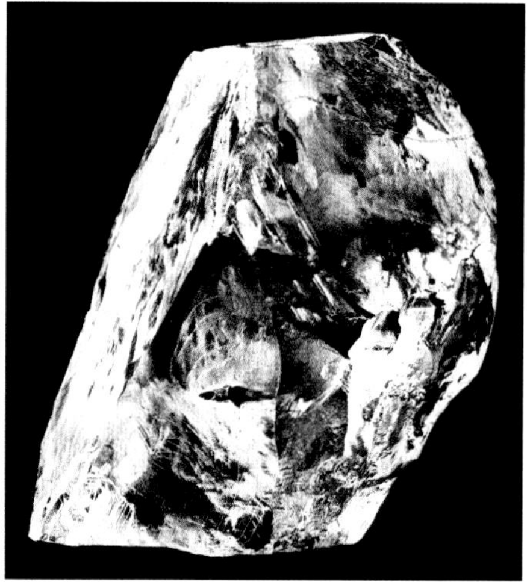

The 3,106.75-carat Cullinan rough diamond and, on the left, the nine Cullinan diamonds cut from it. The diamonds were cut by Joseph Asscher. *Wikimedia Commons*

the major diamond producers. Working together, De Beers and the Syndicate tried to match diamond production to consumer demand.

Maintaining prices by controlling production became more difficult when new mines, such as the Premier, were established. (It was called the Premier mine until 2003, when it was renamed the Cullinan to celebrate its 100th year anniversary and to link the mine to its discoverer and the famous Cullinan diamond.) The mine is located about 300 miles (500 kilometers) northeast of Kimberley and was discovered by Thomas Cullinan in 1902.

The Premier mine became famous when the world's largest gem-quality rough diamond was found at the Premier No. 2 mine in 1905. The stone weighed 3,106.75 carats and was named after the mine's chairman, Thomas Cullinan. In 1907 the Transvaal Colony government bought the diamond and presented it to King Edward VII on his 60th birthday. After the Cullinan was cut, the king had the largest stone, the 530.40-carat Cullinan I, known as the Great Star of Africa, mounted in the Sovereign's Sceptre. Cullinan II, the second-largest stone, is mounted in the Imperial State Crown. Both are part of the British crown jewels. Seven other major diamonds cut from the Cullinan, weighing a total of 208.29 carats, are privately owned by Queen Elizabeth II, who inherited them from her grandmother Queen Mary.

It became impossible for De Beers to own every existing mine, so the company shifted its emphasis to buying rough wherever it was produced in order to safeguard diamond prices and ensure market stability.

## Time Line of South African Diamond Mine Discoveries, Openings and Closures

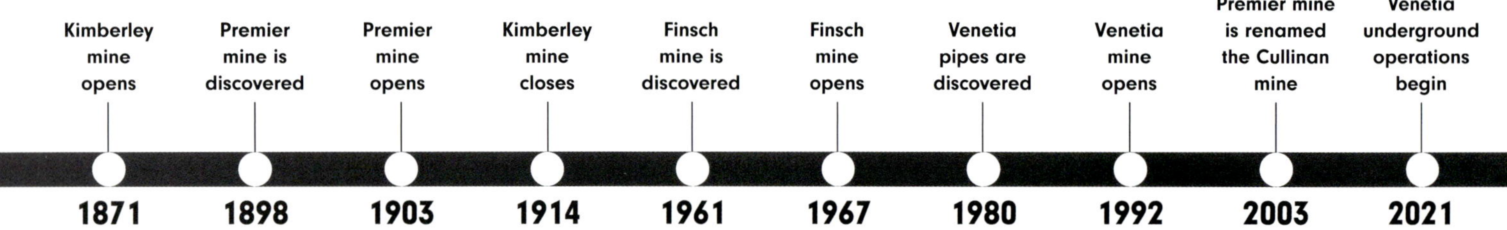


**The Evolution of the London Diamond Syndicate**
In 1930 Oppenheimer created the Diamond Corporation (Dicorp) to succeed the London Diamond Syndicate. In 1934 Dicorp and other major diamond producers joined together to form the Diamond Producers Association (DPA). In 2020 the DPA's name was changed to Natural Diamond Council. As of 2020 it had seven members that represented about 75 percent of the world's rough diamond production: Alrosa, Arctic Canadian Diamond Company, De Beers Group, Lucara Diamond, RZM Murowa Diamonds, Petra Diamonds and Rio Tinto.

In 1929 Ernest Oppenheimer (1880–1957) became the chairman of De Beers, and he remained in control until his death.

During the Great Depression of the 1930s, many small mining companies went out of business. De Beers bought them and gradually gained control of all South African production that was not controlled by the government. When World War II began in 1939, De Beers was the world's major supplier of industrial diamonds, which were needed for military use.

De Beers's trading contracts with the world's diamond producers went largely unchallenged until the mid-1990s. Since then, important producers in Russia, Australia and Canada have taken more control over their own diamonds. In addition, the Cullinan (formerly Premier) mine, which had previously been controlled by De Beers, is now 74 percent owned by Petra Diamonds. Petra continues to mine this underground resource. The Cullinan mine has produced more than

These 424.89-carat and 209.20-carat D-color Type II diamonds were recovered at Cullinan in March and April 2019. *Photo © Petra Diamonds*

The Letlapa Tala collection: Five blue diamonds found at the Cullinan mine in one week in September 2020. *Photo © Petra Diamonds*

750 stones greater than 100 carats and is the world's major source of blue diamonds.

In September 2020 five blue diamonds, ranging from 9 to 25 carats in weight, were found at the Cullinan mine within the space of one week. The collection was named Letlapa Tala, which means "blue rock" in Northern Sotho, the predominant language in the Cullinan area. After a competitive tender process, Petra Diamonds sold the diamonds to a partnership between De Beers and Diacore for $40.36 million.

The previous significant discovery of a blue diamond at the Cullinan mine had been a 20.08-carat blue crystal in September 2019. It sold for $14.9 million to an anonymous buyer.

De Beers also gave up control of the Finsch mine, which is located about 103 miles (165 kilometers) northwest of Kimberley. In 2011 De Beers sold it to Petra Diamonds for $200 million. The shareholding in the Finsch mine now stands as follows: Petra Diamonds (74 percent), Kago Diamonds (Pty.) Ltd. (14 percent) and Itumeleng Petra Diamonds Employee Trust (12 percent).

The Venetia kimberlite mine, located in northeastern South Africa, has been the country's largest diamond producing mine since 1995. It opened in 1992 and is fully owned by De Beers Consolidated Mines. The existing Venetia open-pit mine will be converted to an underground operation by 2021 to prolong the life of the mine to 2046.

South Africa continues to produce diamonds from alluvial deposits within the country, some of which were first mined in the 1800s. Diamond mining also takes place along the coastline of the country in marine deposits.

Specialists at the Finsch mine conduct a safety inspection before beginning production drilling. *Photo © Petra Diamonds*

An evening view of the Finsch mine site in South Africa. *Photo © Petra Diamonds*

# Major African Diamond Deposits

## Namibia

Namibia sits to the northwest of South Africa, along the Atlantic coast. Diamonds were discovered in the country in the sand dunes of the Namib desert in 1908, when Namibia was still known as German South West Africa. The diamonds are believed to have originated from kimberlites in the interior of South Africa. In 1909 almost 500,000 carats were mined, and production kept growing until the beginning of World War I.

In 1928 there were major diamond finds in raised beaches along the Namibian coast north of the mouth of the Orange River. Over decades, mining extended farther and farther out to the ocean, and by the 1960s diamonds were being recovered by submarine mining from the ocean floor near the coast.

Today diamonds are being mined offshore in Namibia's exclusive economic zone. This activity, which takes place in water over 460 feet (140 meters) deep, has made Namibia the world's leading undersea miner. The diamonds produced from these deposits are of exceptional quality because they have been weathered from their source rock in the interior of the African continent, washed down rivers into

Marine mining operations on the Namibian coast. *Photos © Namdeb Diamond Corporation*

The MV *Mafuta*, the world's largest offshore diamond-mining vessel, in Lüderitz Bay, off the coast of Namibia. *Ulrich Doering/Alamy Stock Photo*

the Atlantic Ocean and then transported by waves and longshore currents along the African coast. Only the most durable diamonds could survive all of this travel, so a high percentage of Namibian diamonds are of gem quality with a high average value per carat.

Most of the diamond-mining activity in Namibia is done by Namdeb Diamond Corporation, a partnership owned in equal shares by the government of the Republic of Namibia and De Beers Group. In 2019 Namibia was the world's fifth-largest producer of diamonds in terms of world value.

## Democratic Republic of the Congo (DRC)

A 10.73-carat Fancy yellow diamond crystal from the DRC. *Diamond and photo courtesy of The Arkenstone; photo by Joe Budd*

The Democratic Republic of the Congo (DRC), formerly known as Zaire and often referred to simply as the Congo, is located in Central Africa and became a diamond producer in 1907 after diamonds were found in the Tshikapa region along the Kasai River and its tributaries. A more prolific source of diamonds is the deposit at Mbuji-Mayi (formerly Bakwanga). This city sits on a complex of interlinked diamond pipes that produce mainly industrial-grade diamonds. Many are cubes with an opaque coating. The Mbuji-Mayi diamond field is so large that it extends into the neighboring country of Angola. Higher-quality diamonds are found in the alluvial deposits west of the city.

The DRC was the world's sixth-largest diamond producer in 2020. The majority of the diamonds are industrial grade, and most of the DRC's diamond production is done by artisanal miners and not mining companies. This has made it easier for illegal smuggling to occur in the country, which has suffered much political instability. The only commercial diamond producer in the country is Societé Minière de Bakwanga (MIBA), a joint venture between the Belgian company Sibeka and the DRC government. Approximately 80 percent of MIBA's stock is owned by the Congolese government, with 20 percent owned by Sibeka, which in turn is owned by Mwana Africa PLC, which operates the mines.

Angola's largest recorded diamond at 404.20 carats, the 4 de Fevereiro (February 4th) stone was recovered at Lulo in February 2016 and sold for $16 million. *Photo © Lucapa Diamond Company Limited*

## ANGOLA

Angola sits along the Atlantic coast of Africa on the northern border of Namibia and south of the DRC. Diamond mining began in Angola in 1916 while it was a Portuguese colony. The country became a major source in 1921, when 100,000 carats were produced from its alluvial deposits. Most were found in the province of Lunda Norte in the northeastern part of the country.

The first kimberlite pipes were found in northeastern Angola in 1952, and since then many more have been found. The largest is Catoca, in the eastern province of Lunda Sul, close to the city of Saurimo. The kimberlite mine began operation in 1997 and has become the world's fourth-largest diamond mine. It is operated by Sociedade Mineira de Catoca, a joint venture between Angola's state-owned diamond company Endiama, Russian-owned Alrosa and China-based company Lev Leviev International.

Angola is especially noted for its Lulo alluvial mine in Lunda Norte because of the high average dollar value of its rough. The diamonds recovered from the mine include large high-quality diamonds weighing more than 10 carats, high-end Type IIa diamonds, and pink- and yellow-colored diamonds. Type IIa diamonds are rare and often colorless because they contain very little nitrogen.

The 15.20-carat heart cut from 46-carat Fancy Intense orangy-pink rough found at the Lulo mine in 2020. *Photo © Lucapa Diamond Company Limited*

In February 2016 a 404.20-carat Type IIa diamond was recovered at the Lulo mine. It is the biggest diamond ever found in Angola and sold for $16 million. A year later a 227-carat Type IIa D-color diamond was also found.

The Lulo mine is operated by the Sociedade Mineira do Lulo and is owned by the Australian-based Lucapa Diamond Company and two Angolan partners: Endiama and Rosas & Pétalas SA.

In the fall of 2020 Lucapa unveiled a 15.20-carat heart-shaped pink diamond that was cut and polished from a 46-carat rough from the Lulo mine. GIA graded it as a Fancy Intense orangy-pink with a clarity grade of VVS1. Two pear-shaped pink diamonds weighing 3.30 carats and 2.30 carats, respectively, were also cut and polished from the rough. According to Lucapa, the 46-carat pink rough was the largest gem-quality fancy color diamond recovered to date from the Lulo mining operations.

## Sierra Leone

Sierra Leone is a country on the southwestern coast of West Africa. It is bordered by Liberia to the southeast and Guinea to the northeast.

Alluvial diamonds were discovered in western Sierra Leone in 1930 in the Yengema-Koidu region. It was in the Diminco alluvial mines that the 968.90-carat Star of Sierra Leone diamond was found in 1972. It was purchased by New York City jeweler Harry Winston. The diamond was initially faceted as an emerald-cut stone weighing 143.20 carats, but because of an internal flaw it was later recut into 17 smaller stones, of which 13 were flawless. The largest was a flawless pear-shaped diamond of 53.96 carats. Winston later set six of the diamonds, including the largest one, into the Star of Sierra Leone brooch.

The first kimberlite pipes in Sierra Leone were found in 1948. Since 2003 two kimberlite pipes in the Kono district have been operated by Koidu Holdings SA, which is wholly owned by BSG Resources Limited. The mine's area is referred to as the Koidu Kimberlite Project.

High-quality diamond crystals from the Williamson mine in Tanzania. *Photo © Petra Diamonds*

## TANZANIA

Tanzania is on the eastern coast of Africa, south of Kenya and north of Mozambique. In 1940 the first kimberlite mine outside of South Africa was established there by Canadian geologist Dr. John Williamson and is appropriately called thc Williamson (Mwadui) diamond mine. It holds the record for being the world's longest-running uninterrupted diamond mining operation.

The mine is renowned for its pink diamonds, including the 54.50-carat Williamson Pink, which is considered one of the finest pink diamonds ever recovered. After the diamond was found in 1947, Williamson presented it to Princess Elizabeth as a wedding present. It was cut into a flawless 23.60-carat round brilliant and set in a flower brooch designed by Cartier.

The Williamson mine is 75 percent owned by Petra Diamonds and 25 percent owned by the government of the United Republic of Tanzania. Petra plans for the mine to operate until at least 2033.

A 23.16-carat pink diamond found at the Williamson mine in November 2015. *Photo © Petra Diamonds*

## Botswana

Botswana is bordered by Namibia on the west, South Africa on the south and Zimbabwe on the northeast. In 2020 it was the world's second-largest producer of diamonds. After years of exploration, De Beers discovered the Orapa kimberlite pipe in 1967. At 292 acres (118 hectares), it was the biggest pipe De Beers had ever found. As of 2020 it is the world's fifth-largest diamond mine in terms of measurable reserves. The Orapa mine is owned by Debswana, a 50/50 partnership between De Beers and the government of Botswana.

The Jwaneng diamond mine in south-central Botswana is also owned by Debswana. It began operations in 1982 and is the world's second-largest diamond mine in terms of reserves and the richest diamond mine in the world by value. It contributes 60 to 70 percent of Debswana's total revenue. A major extension project at Jwaneng has extended the mine's life to at least 2034.

Debswana, which also owns the Letlhakane and Damtshaa mines in Botswana, is the biggest contributor to De Beers Group's rough diamond production. The company is committed to mining safely and responsibly as well as making a meaningful contribution to the development of communities around its mines and the nation at large. Botswana is a stable multiparty democracy that has used its diamond profits to improve living standards for all of its citizens.

The 1,109-carat Lesedi La Rona.
*Photo © Lucara Diamond*

The Karowe mine in Botswana is another noted producer of high-quality diamonds. It is fully owned by Lucara Diamond Corporation, a Canadian diamond mining company with exploration licenses in Botswana. A 1,109-carat Type IIa diamond named Lesedi La Rona was discovered at the Karowe mine in 2015. It is the world's second-largest single-crystal colorless rough diamond and was sold to London jeweler Laurence Graff for $53 million. It was cut into the world's largest square emerald-cut diamond, a 302.37-carat D-color diamond named the Graff Lesedi La Rona.

In addition to the main diamond, 66 other diamonds were polished from the rough, ranging in weight from under a carat to more than 26 carats. Each diamond is inscribed with "Graff Lesedi La Rona" and its unique GIA number. According to GIA, the Lesedi La Rona rough is a "superdeep" diamond formed three times deeper than most other diamonds. Graff donated fragments of the Lesedi La Rona to the Smithsonian Institute to help advance diamond research.

The 1,758-carat Sewelô diamond.
*Photo © Lucara Diamond*

In April 2019 the Sewelô diamond was found at Karowe. It weighed 1,758 carats and is second only to the historic 3,107-carat Cullinan Diamond recovered in South Africa in 1905. The Sewelô has a dark gray appearance on its surface and was sold to Louis Vuitton at the beginning of 2020.

## Zimbabwe

Formerly known as Rhodesia, Zimbabwe is a landlocked country west of Mozambique and north of Botswana and South Africa. In June 2006 the Marange diamond fields were discovered in the Mutare district in eastern Zimbabwe. Considered one of the world's largest alluvial diamond deposits, Marange is an area of widespread small-scale diamond production. It has been controversial because of looting and illegal mining that occurred there in the past.

An open-pit kimberlite diamond mine named Murowa is located in Mazvihwa, in south-central Zimbabwe. It began production in 2004 and is operated by RioZim.

## Lesotho

Diamond rough from the Mothae mine. *Photo © Lucapa Diamond Company Limited*

Lesotho sits within the borders of South Africa. It was previously the British Crown Colony of Basutoland, but it declared independence from Great Britain in 1966. A new democratic government was elected in 2017.

The world's highest diamond mine (which sits 10,171 feet or 3,100 meters above sea level) is Letšeng-la-Terai in northeastern Lesotho. The mine is owned by Gem Diamonds and the government of Lesotho. The Letšeng diamond mine is renowned for its production of large diamonds of exceptional quality, making it the highest average price per carat kimberlite diamond mine in the world. Its famous diamonds include the 910-carat Type IIa D-color Lesotho Legend, the 603-carat Lesotho Promise, the 550-carat Letšeng Star and the 493-carat Letšeng Legacy. A 442-carat Type II diamond was recovered in August 2020. Letšeng has also produced high-quality pink and blue diamonds.

Another high-value kimberlite mine called Mothae is located within 3 miles (5 kilometers) of Letšeng. The Mothae mine is 30 percent owned by the Lesotho government and 70 percent by Lucapa Diamond Company, the same company that owns the Lulo alluvial mine in Angola. Mothae started commercial operations in December 2018. The first phase of the mine targeted the weathered kimberlite

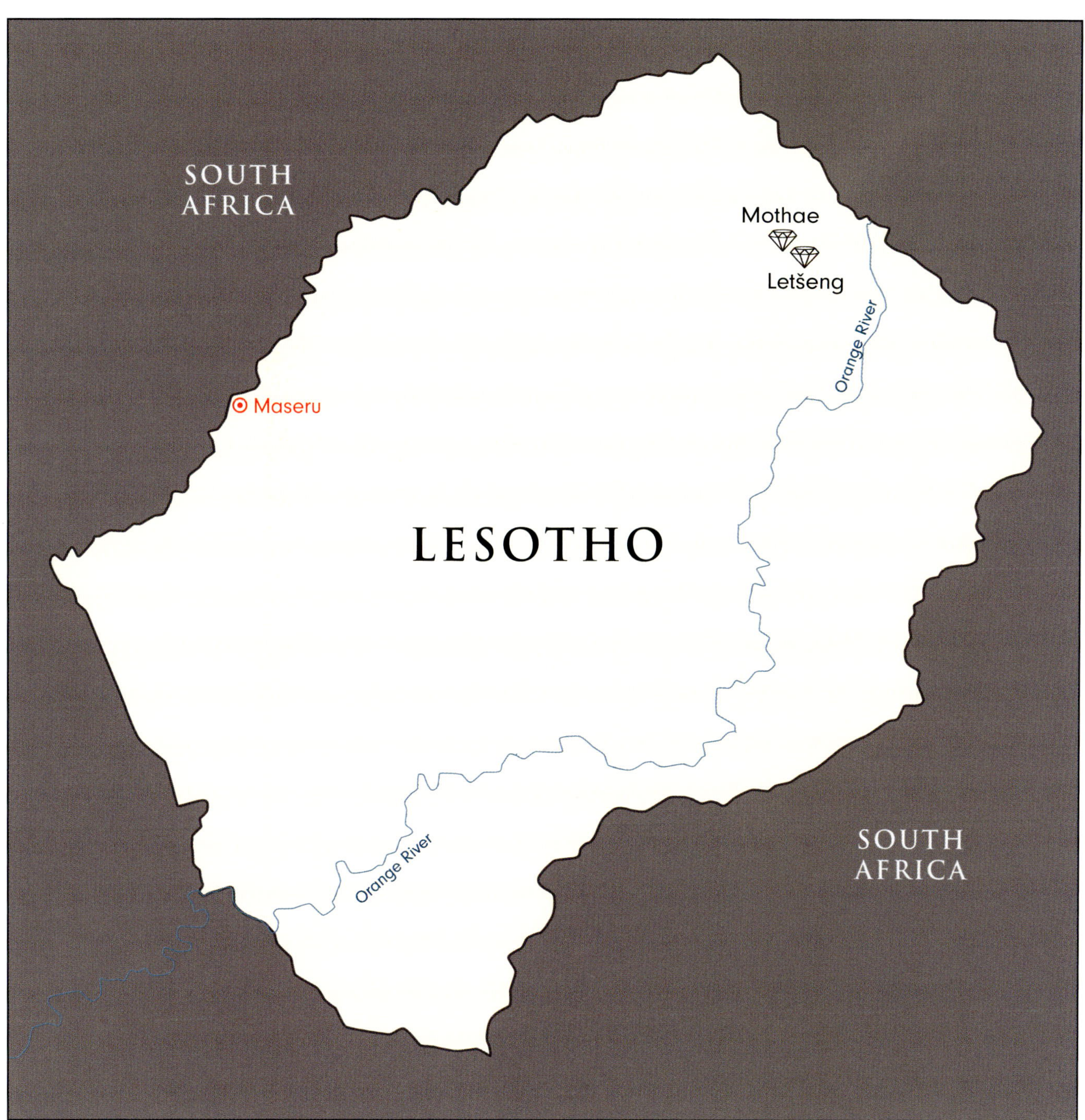


material present at the top, while the second phase (projected to start in 2021) will involve mining the harder and unweathered kimberlite below the pipe. The extracted kimberlite resources are processed at the on-site treatment plant that houses two X-ray transmission diamond recovery modules, which are capable of recovering Type IIa diamonds out of the secondary crushing circuit.

A park ranger helps a young prospector search for diamonds. *Arkansas Department of Parks, Heritage and Tourism*

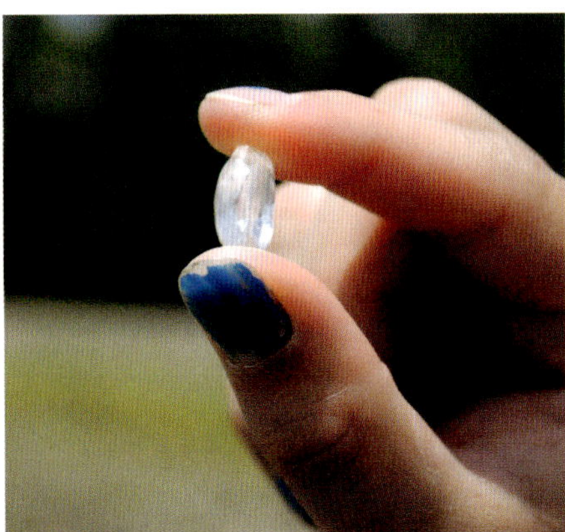

The 8.52-carat Esperanza diamond in its rough form. *Arkansas Department of Parks, Heritage and Tourism*

## USA

The Crater of Diamonds State Park in Arkansas is the only diamond-bearing site in the world where visitors can look for diamonds and keep what they find.

The first diamond was found here in 1906 by John Huddleston, when the land was his farm. This led to a search of the surrounding countryside, where four pipes were discovered, two of which have produced diamonds. Huddleston sold the farm to a group of Little Rock investors, but commercial attempts at diamond mining were not profitable. In 1972 the state of Arkansas bought the land for $750,000 and turned it into a state park and tourist attraction. More than 33,000 diamonds have been found by visitors since the park opened. In 2015 a lady who had paid only $8 to search for diamonds at the park found an 8.52-carat colorless diamond and named it the Esperanza. The crystal was cut into a 4.60-carat Type IIa D-color internally flawless diamond.

Between June 1996 and late 1997 diamonds were commercially mined in Colorado near the Wyoming border at the Kelsey mine. The area consists of nine kimberlite pipes, two of which were open-pit mined but are now dormant. Diamond mining is costly, and if it is not economically profitable to extract diamonds from the ground, they will remain there.

Alluvial diamonds have been found in several states, including California, North Carolina, Oregon, Virginia, West Virginia and Wyoming. According to a 1994 report entitled *Pacific Coast Diamonds – An Unconventional Source Terrane* by W. Dan Hausel, diamonds were first discovered in California in 1849 during the Gold Rush by prospectors panning for gold near Placerville, east of Sacramento. Diamonds are still found in streams and rivers of the Sierra Nevada mountains and even along the Pacific coast, but not in commercial quantities.

## RUSSIA

Russia is the world's largest producer of diamonds both in terms of volume and value.

Alluvial diamonds have been reported in Russia since 1829, but intensive exploration did not start until 1947.

In August 1954, the first Russian diamond pipe was discovered in the northeastern Siberian Republic of Sakha (also known as Yakutia). The pipe was named Zarnitsa, meaning "summer lightning." In June the next year, the Mir ("Peace") pipe was found about 373 miles (600 kilometers) south of Zarnitsa and was followed by the discovery of the Udachnaya ("Lucky") pipe two days later, just 11 miles (17.5 kilometers) west of Zarnitsa.

The first commercial diamond mine to open in Russia was the Mir mine in 1957. This was a major achievement because for about seven months of the year, temperatures in Yakutia can be as low as −85°F (−65°C) and the location is almost inaccessible. Raw materials, equipment, food and other supplies can only be shipped into the area during the brief summer season. Otherwise they have to be brought in by plane. The city that grew around the Mir mine is called Mirny and has a population of about 40,000.

Most Russian diamond mines today are owned by Alrosa, a Russian group of diamond mining companies that specialize in

Machines remove layers of dirt, gravel and rock from the open-pit mine at the Nyurbinskaya pipe in Yakutia, Russia. *Photo © Alrosa*

### Time Line of Russian Diamond Mine Discoveries, Openings and Closures

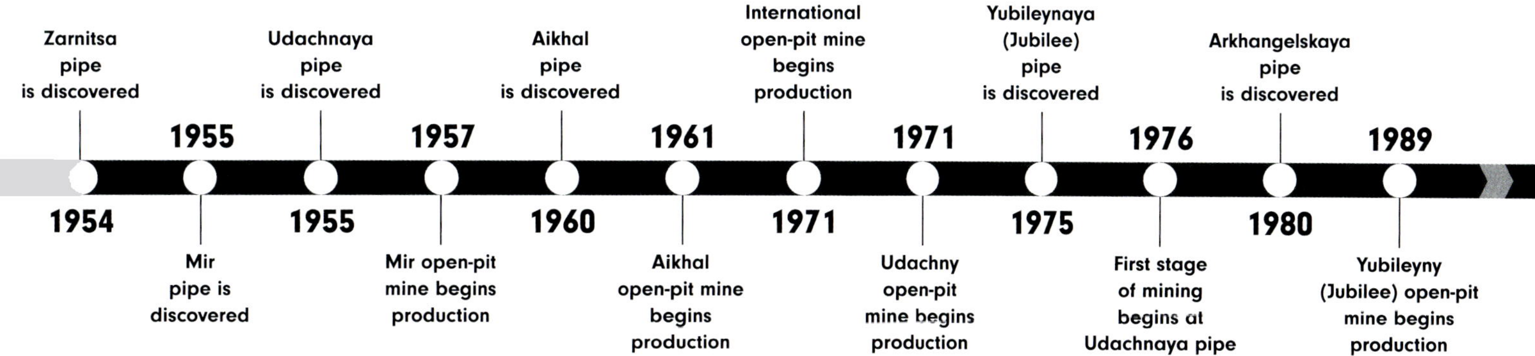

A road spirals around the open-pit mine at the Nyurbinskaya pipe, allowing trucks to drive in and out of the central mining area. *Photo © Alrosa*

the exploration, mining, manufacture and sale of diamonds. It was founded in February 1992 a few months after the dissolution of the Soviet Union. About a year later, its full name — Almazi-Rossii-Sakha — was shortened to Alrosa, and a subsidiary named Severalmaz ("Northern Diamond") was created to oversee exploration in the Arkhangelsk region. In 2011 Alrosa was reorganized as an open

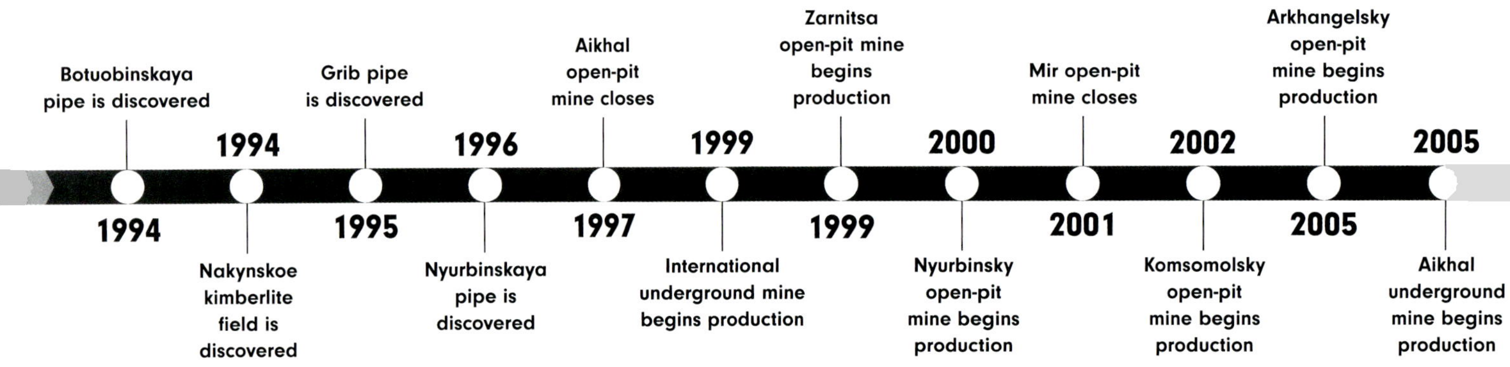

public joint-stock company with free float of Alrosa's shares on financial markets. The main headquarters for Alrosa are in Mirny and Moscow, but it has offices throughout the world.

Alrosa's main mining and processing operations are located in two Russian regions: the Republic of Sakha (Yakutia) and the Arkhangelsk region in the northwest of the Russian Federation. The company has open-pit and underground mining as well as alluvial operations in Sakha and open-pit mines in the Arkhangelsk region.

Alrosa's 11 primary and 13 alluvial deposits and its processing centers are grouped by territory into four mining and processing divisions (MPDs) — Mirny, Udachny, Aikhal and Nyurba — and two subsidiaries — Almazy Anabara (including mining and processing capacities of Nizhne-Lenskoye) and Severalmaz. Each MPD consists of several mines, processing plants and stocks of equipment.

Another diamond pipe found in the northwestern part of Russia is the Grib pipe. It was discovered in 1995, and a mine was opened in 2014. Named after Vladimir Grib, one of the Russian geologists who discovered diamonds in the area, the mine is privately owned by the Russian financial group Otkritie Holding. Grib is operated by AGD Diamonds and its diamonds are marketed by Grib Diamonds in Antwerp. Located about 80 miles (130 kilometers) from the regional capital of Arkhangelsk, the Grib mine is one of the largest diamond mines in the world.

Russia is the world's chief source of colorless to near colorless diamonds, and it is expected to become the leading supplier of fancy

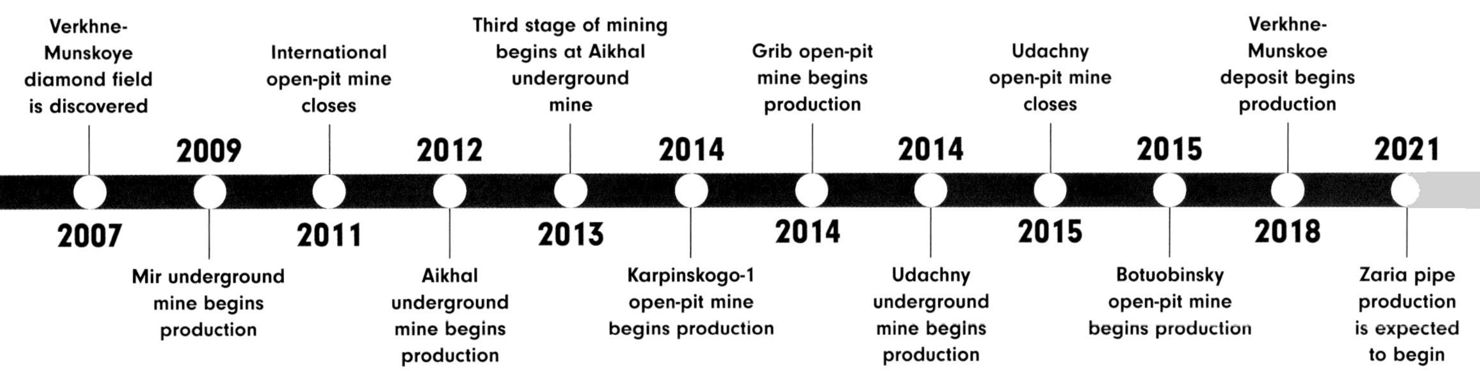

ARKHANGELSK OBLAST
Moscow
REPUBLIC OF SAKHA (YAKUTIA)
RUSSIA

A 27.85-carat rough pink diamond found in October 2017 at the Ebelyakh alluvial deposit in Sakha. It was cut into a 14.83-carat oval diamond named the Spirit of the Rose, shown on the left. *Photos © Alrosa*

color diamonds now that the Argyle mine in Australia has closed. Lomonosov MBD in northwestern Russia has produced a significant number of yellow, pink, purple, brown and green diamonds.

Colored diamonds are also found in Russia's Yakutia mines, and they can be quite large. Two notable examples come from the Ebelyakh alluvial deposit. One is a 236-carat yellow-brown diamond found in August 2020. The other is a 27.85-carat rough pink diamond found in October 2017. The latter was cut into a 14.83-carat internally flawless Fancy Vivid purple-pink diamond named the Spirit of the Rose. The process of cutting and polishing it took almost a full year. In November 2020 it sold for $26.6 million at a Sotheby's Geneva auction and set a record for the highest price ever paid at auction for a Fancy Vivid purple-pink diamond.

Besides being noted for its large volume of high-quality diamonds, Alrosa is also respected for its tradition of social responsibility. It creates high-paying jobs, provides funding for social services and maintains high health, safety and environmental standards in the remote areas of Russia where it operates.

An Argyle pink diamond in a ring made by Gems by Pancis. *Photo from Gems by Pancis*

An Argyle brown diamond macle. *Crystal and photo courtesy of The Arkenstone*

A Fancy Vivid purplish-pink Argyle diamond. *Photo courtesy of Joe Namdar of Namdar Diamonds*

## AUSTRALIA

Alluvial diamonds have been found in the states of New South Wales since 1851 and Western Australia since 1895. Diamond exploration began in Western Australia in 1972, and by 1976 the first lamproite pipe was discovered in the Ellendale area of Western Australia. Three years later another pipe, now called the Argyle mine, was found in Kimberley — the northernmost region of Western Australia. The mine was established in 1983 as the first major diamond operation in Australia, and by 1985 it was the world's most prolific diamond mine in terms of volume. In its peak year, 1994, it produced 40 percent of the world's diamonds.

According to the Spring 2001 issue of *Gems & Gemology*, the average weight for Argyle rough was less than 0.10 carats, but crystals as heavy as 42 carats were found. More than 60 percent of Argyle crystals were irregular in shape; triangular macles comprised about 25 percent of the mine's output. About 72 percent of the diamonds were brown, and most of the remaining stones were yellow to near-colorless or colorless. The mine's owner, Rio Tinto, successfully marketed the brown to light-brown diamonds as "cognac" and "champagne" colored. India became the major cutting center for these diamonds, which found their way into budget-priced jewelry throughout the world.

Less than 1 percent of Argyle's diamonds had a rare pink, red, grayish-blue or green color. However, the number of pink diamonds produced was significant enough for Argyle to become famous for its pink diamonds. Argyle used four principal categories — pink, purplish pink, brownish pink and pink "champagne" — to grade its pink stones. The color intensity was graded from very faint to very intense.

Rio Tinto closed the Argyle mine in late 2020, but long before it began the project it had worked with the government, Indigenous groups and local communities to plan the closure of the mine in a way that would bring lasting benefits to the host community. For example, the company has been restoring and replanting the land at the mine, and it has provided career support and training to help the local workforce transition to new jobs. Following the last production of diamonds, Rio Tinto anticipated that it would take five years to decommission and dismantle the mine and undertake rehabilitation, followed by a further period of time for monitoring, during which Argyle would continue to employ people post-mining.

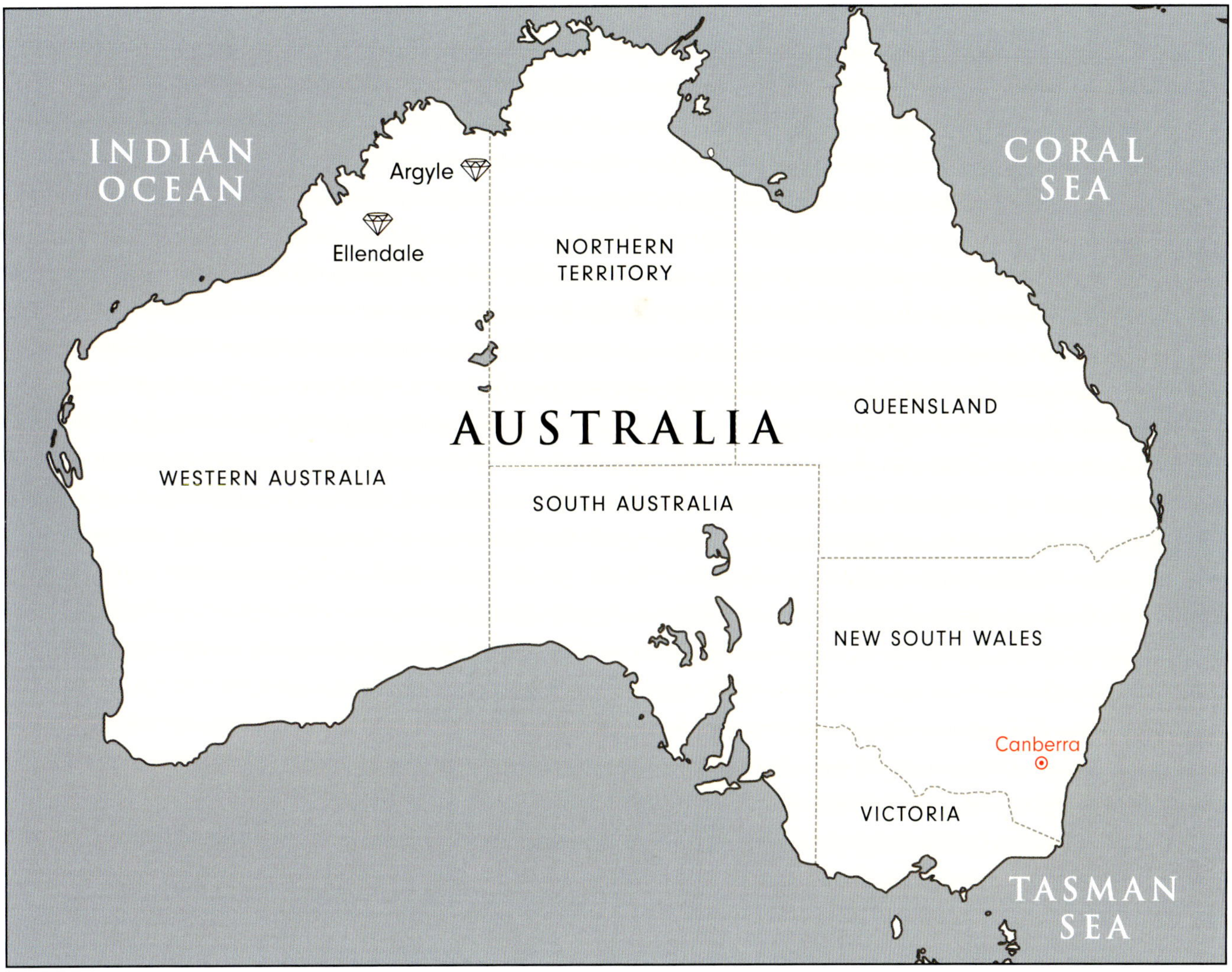

About 248 miles (400 kilometers) southwest of Argyle is the Ellendale mine, which was famous for its yellow diamonds. It opened in 2002 and closed in 2015. Two years later India Bore Diamond Holdings (IBDH), a privately held Australian mining company, began a drilling and trenching program in the area to search for ancient alluvial gravel beds containing diamonds buried more than 49 feet (15 meters) below the surface.

In August 2020 IBDH announced that it had recently discovered a significant alluvial deposit of rare yellow diamonds at Ellendale. Many of the stones displayed a distinctive purple fluorescence under UV light. It is hoped that renewed exploration at Ellendale can provide a new source of colored diamonds for Australia, as well as employment.

Ten Argyle pink diamonds surrounding a Fancy Intense pink diamond. *Ring and photo courtesy of Brian Denney of Gems of Note*

## CANADA

Canada's first diamond mine opened in 1998. The Ekati mine is in a remote area of Arctic tundra in the Northwest Territories, about 186 miles (300 kilometers) northeast of Yellowknife, the nearest supply center, and about 124 miles (200 kilometers) south of the Arctic Circle. The only access to this area is by air or by roads that are passable only when the ground is frozen solid. Nevertheless, mining and processing occur here 24 hours a day and 365 days a year (except during the COVID-19 pandemic).

Before the Canadian government would issue permits to develop the property, the Australian exploration and mining company Broken Hill Proprietary Co. (BHP) had to do an environmental study proving it would not endanger the region's natural resources. The Canadian government also required BHP to involve Canadian workers in

An aerial view of the Ekati mine site in the winter. *HeavilyMeditated/ Shutterstock*

ARCTIC OCEAN
GREENLAND
NORTHWEST TERRITORIES
Jericho
NUNAVUT
Ekati & Diavik
Snap Lake
Yellowknife
Gahcho Kue
Chidliak
Iqaluit
CANADA
ALBERTA
SASKATCHEWAN
MANITOBA
Renard
Victor
QUEBEC
ONTARIO
Montreal
Ottawa
Toronto
U.S.A.

## Time Line of Canadian Diamond Mine Discoveries, Openings and Closures

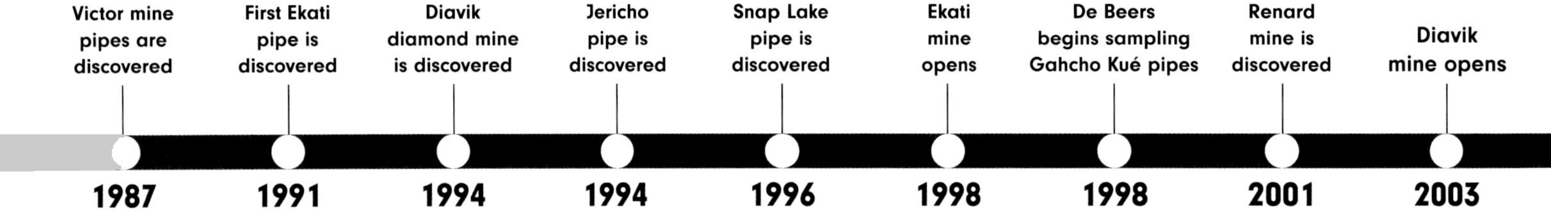


sorting, grading and polishing the diamonds mined there. In 2013 BHP sold the mine to Dominion Diamond Company, who in February 2021 sold it to Arctic Canadian Diamond Company Ltd. Mining operations are expected to continue until at least 2030.

The second diamond mine to open in Canada was Diavik in 2003. It consists of four kimberlite pipes and is located just 19 miles (30 kilometers) southwest of Ekati. The Diavik diamond mine is managed by Rio Tinto and is owned by a joint venture between Diavik Diamond Mines (2012) Inc., which is a wholly owned subsidiary of Rio Tinto (60 percent), and Arctic Canadian Diamond Company Ltd. (40 percent). Mining operations at Diavik are expected to continue until 2024.

According to the Summer 2016 issue of *Gems & Gemology*, the Diavik kimberlite pipes contain exceptionally high grades of moderate- to high-value diamonds. Colorless crystals predominate, though brown and occasionally yellow diamonds also occur. In October 2018 a 552-carat yellow diamond the size of a chicken egg was found at the Diavik mine and captured the record as the largest diamond found in North America.

The Snap Lake mine is less than 62 miles (100 kilometers) south of Diavik and Ekati. De Beers began production at Snap Lake in January 2008 but closed the mine in 2015 because it had water issues and was not profitable. It was Canada's first completely underground diamond mine and produced 7.8 million carats of diamonds.

In July 2008 De Beers opened the Victor mine in northern Ontario. The largest diamond found at Victor weighed 271 carats and was cut into a 102.39-carat D-color flawless oval diamond. In October 2020 it was sold online at a Sotheby's Hong Kong auction for $15.68 million, beating the previous online diamond auction record of $2.1 million set in July 2020 for a 28.86-carat diamond. In addition to setting an online record, the diamond was also the most expensive Canadian

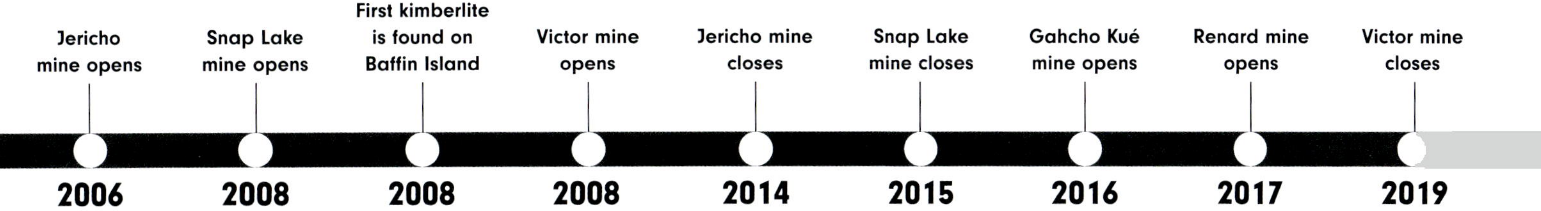


diamond ever sold in any manner. The Japanese buyer of the diamond named it the Maiko Star after his second daughter.

More than 8.1 million carats of diamonds had been recovered at Victor when production ended there in May 2019. Through the end of 2020 almost 40 percent of the mine site had been restored with more than 1.2 million trees being planted since progressive reclamation began in 2014. Reclamation of the mine site is expected to be complete by 2024.

East-central Quebec is another source of Canadian diamonds — the Renard mine. It began commercial production in January 2017 and is 100 percent owned by the Stornoway Diamond Corporation. Operations at Renard were suspended for six months in 2020 as a result of the COVID-19 pandemic. It is expected to have a mine life of 25 years.

In 2008 the first kimberlite was discovered on Baffin Island, Nunavut, during exploration conducted by BHP Billiton and Peregrine Diamonds. Since then, more than 70 kimberlites have been found and are referred to as the Chidliak Project (or simply Chidliak). In 2011 BHP sold its 51 percent interest in the project to its partner, Peregrine Diamonds. Then in September 2018 Peregrine sold Chidliak to De Beers Canada.

Diamonds from Canada have been very popular with Canadians and people who are concerned about environmental and human rights issues. Many diamonds mined in Canada are graded and have their diamond report number laser inscribed on their girdle along with a trade logo, such as a maple leaf, polar bear or Canadamark. This assures buyers that their diamonds are truly Canadian and that the quality of the diamond corresponds to the grades on the lab report. Canadian diamonds are generally of high quality and, as a result, command a high average price per carat.

The Chidliak Project's Discovery Camp in 2019. *Photo © De Beers Group*

# 3 Diamond Mining and Processing

The first diamonds were mined by hand in riverbeds and streambeds, and some diamonds are still mined this way. However, new methods have also been developed and depend on the location of the diamonds. This chapter will discuss four methods of diamond mining: alluvial mining, marine mining, open-pit mining and underground mining.

## Alluvial Mining

Before the discovery of kimberlite diamonds pipes in South Africa in 1869, all diamonds were found in alluvium — the silt, sand and gravel of deltas, dried-up riverbeds, and active rivers and streams. These diamonds are called alluvial diamonds, and the mining method is referred to as alluvial mining. This method is divided into two types: wet digging and dry digging.

Wet digging (a.k.a. panning) is done while washing the sediment and small rocks collected from a river. This technique was adopted from gold miners, and records of this method go back as far as ancient Rome.

First, the miner identifies areas where diamonds might be in the gravel on the riverbed. Then they fill a large pan with gravel and water and shake it back and forth to settle the heavy material to the bottom. The lighter material washes over the top of the pan, and the miner removes the larger rocks. These steps are repeated until diamonds and/or other gemstones may be visible in the sediment left in the pan.

Sometimes miners dive for diamonds by collecting sediment from the bottom of a river and then panning it at the surface. Miners also

A miner uses a large sieve to look for diamonds in the Kono district of Sierra Leone. *Des Willie/Alamy Stock Photo*

wash river gravel in large baskets and circular sieves. Swirling the sieves allows the smaller grains to escape while the mesh traps larger, heavier stones. The miners then pick through the remaining material by hand.

Wet digging is still done in countries like Sierra Leone, Angola, Zimbabwe, Indonesia and the Democratic Republic of the Congo.

Dry digging is done in dried-up riverbeds and can involve blocking the flow of a river at both ends using dams to create a dry area where miners can then collect gemstones. The mining can be done by hand with picks, shovels and sieves or using larger machinery, like bulldozers, anywhere there is a dried-up riverbed containing diamonds.

Not all alluvial mining is done on a smaller scale. In larger operations, recovery and sieving are mechanized. In some cases, miners use large equipment to transport the diamond-bearing earth to a processing plant. A series of sieves separates the diamonds from the surrounding waste material before the diamonds are loaded into a washing pan. Final processing may include a moving grease belt or grease table. While on the greased surface, water is sprayed over the diamonds and other materials. Diamonds are naturally drawn to grease and repel water, so they stick to the greased surface while the remaining minerals and dirt are washed away. Afterward the rough alluvial diamonds are sorted into various categories.

Artisanal divers search for diamonds at the bottom of the Sewa River in Sierra Leone. *Photo by Laurent Cartier*

An artisanal miner in Sierra Leone captures a digital record of the diamond he found. *Photo © De Beers GemFair*

Tracking diamonds from countries with alluvial mining and verifying that they are conflict free is a challenge, especially when the mining is done by independent individuals referred to as artisanal miners. Nevertheless, these miners deserve an opportunity to make a living from the diamonds they find, and some companies are stepping up to assist.

To help artisanal miners get fair treatment and ensure that their diamonds are sold ethically, De Beers has started a program called GemFair in Sierra Leone to connect artisanal and small-scale miners to the global market through digital technology and assurance of ethical working standards. Miners are trained by GemFair on ways they can improve their work and business practices, and they have access to tutorials on the program's app, which was prepared by leading diamond experts. Miners receive fair value for their diamonds, and GemFair helps them create a digital record of the ones they have found at registered mine sites. This allows their diamonds to be traceable from mine to market. If the pilot program works well in Sierra Leone, De Beers hopes to expand it to other countries.

## Marine Mining

Diamond mining areas that are on an ocean shoreline or offshore are called marine deposits. The largest ones are located off the Atlantic coasts of Namibia and South Africa, north and south of the Orange River. The deposits of diamonds found there are a result of kimberlite pipe erosion. When heavy rain occurs, exposed diamonds are washed into rivers and carried toward the coast. Four main factors affect the distribution of marine deposits:

- **Sea level fluctuations:** Over the last 100 million years, the sea level has fluctuated from more than 1,600 feet (500 meters) below present levels to more than 980 feet (300 meters) above them. This means that diamonds have been deposited at several elevations above and below the current sea level.
- **Ocean currents:** Ocean currents have pushed sand and diamonds northward from the mouth of the Orange River, and at the same time the currents have formed a "wall" that has prevented the diamonds from being pushed farther out to sea.
- **The size of the diamonds:** The smaller and lighter the diamonds, the farther they can be carried out to sea. Near the mouth of the Orange River, the average diamond size is 1.50 carats, but 125 miles (200 kilometers) north of the river the average is 0.10 to 0.20 carats.
- **Ocean waves:** The pounding action of the surf can force diamonds into small cracks and crevices along the shoreline. Dislodging and removing these diamonds can require special tools and procedures. Fractured diamonds that survive the trip down the Orange River are usually broken apart by the waves along the shoreline, leaving mainly fracture-free diamonds. About 90 to 95 percent of Namibia's diamonds are gem quality.

Marine diamonds are found in a variety of areas and conditions, so mining companies use several methods to mine them, including modified alluvial mining, shallow-water mining using divers, mining with huge vacuums that pick up the sand and transport it to a processing plant and deep-sea operations involving high-tech ships.

Marine mining operations on the Namibian coast. Sometimes workers use tremendous vacuums to carry loosened ore material directly to a processing plant. *Photos © Namdeb Diamond Corporation*

Deep-sea mining takes place on the ocean floor. The diamonds are usually extracted using a hydraulic suction system or a continuous-line bucket system. The latter is the preferred method and operates like a conveyor belt running from the seafloor to the surface, where the ship or mining site extracts the desired gems and returns the leftover sediment back to the ocean.

## Open-Pit Mining

Open-pit mining is a technique for extracting diamonds from diamond pipes in the Earth's crust by digging, eventually creating a pit. This is the method initially used when mining diamond pipes. Before the mine opens, however, it must be located and developed. This can take years and require millions of dollars.

For example, it took almost 10 years for prospectors Chuck Fipke and Stewart Blusson to discover the first diamond-bearing kimberlite at Canada's Ekati deposit, in 1991. About another seven years were required to develop the mine and do environmental studies to prove to the Canadian government that the mine would not endanger the region's natural resources. The Ekati mine was finally opened in 1998 and became Canada's first diamond mine.

Trucks remove layers of dirt, gravel and rock covering the kimberlite pipe. *Photo © Alrosa*

The digging of an open-pit mine typically begins by removing layers of overburden — the dirt, sand, gravel or rock that covers the pipe. After removal, the overburden is processed to extract any diamonds it might contain. Then the ore in the diamond pipe is blasted loose with explosives. Afterward, the loosened material is removed with hydraulic shovels and large ore trucks. As the open-pit mine develops, it forms a downward-pointing cone that follows the contours of the pipe. The sides are terraced to help prevent rocks from falling and to provide a roadway around the mine's perimeter for the ore trucks to remove material from the pit. The trucks carry the rock to the processing facility, where the diamonds are separated and recovered from the ore.

Many governments require that open-pit mines be returned to their original state after closure. The mining company Rio Tinto had begun restoring and replanting the land at the Argyle diamond mine

A helicopter view of the open-pit mine at the Nyurbinskaya pipe in Yakutia, Russia. Levels form circular rings and are arranged in a spiral to allow trucks to drive in and out of the central mining area. *Photo © Alrosa*

Explosives loosen the ore in the Nyurbinskaya pipe. *Photo © Alrosa*

## Underground Mining

The open-pit Udachny mine in Russia is 2,067 feet (630 meters) deep. It is extremely expensive to remove rock from such depths to access new ore. Consequently, it becomes more economical to transition to an underground mine.

A tunneling machine, which is used to create tunnels in an underground mine. *Photo © Alrosa*

An underground mine consists of a series of vertical shafts and horizontal tunnels in and around the diamond pipe. The vertical shafts are drilled into the stable rock near the pipe to allow access for personnel, supplies, equipment and ventilation. Then horizontal tunnels are drilled from the shafts through the pipe to extract the kimberlite. Tunneling machines can also be used. The ore is loaded into the cars of an electric train that dumps it onto an underground crusher. From the crusher, a conveyor system carries the ore to buckets that take it up the vertical shaft for further processing.

After the kimberlite ore leaves the mine, it goes through a diamond extraction process called recovery. The ore starts out as fairly large chunks that must be reduced in size by a series of crushing and scrubbing operations. From the primary crushing plant, the ore goes to a machine called a scrubber that removes dirt and particles of clay. The ore then passes over a screen. Gravel larger than the screen's holes goes to another crushing plant to further reduce the size. Smaller gravel that falls through the screen moves to the next step, called dense media separation, which involves using ferrosilicon powders to form a suspension in water that has a higher density than water. The gravel is fed into the suspension, and the diamond-bearing gravel, which is heavier, sinks to the bottom and is removed.

After the dense media separation process, the diamonds are extracted from the gravel using a grease belt, grease table and/or X-ray machine. Many diamonds luminesce under X-rays and can therefore be identified and separated in final recovery. As a thin stream of ore passes in front of an intense X-ray beam, the fluorescent light emitted by a diamond crystal triggers a jet of air that blows the stone away from the ore and into a collector bin.

However, some diamonds, especially more valuable Type II stones, do not luminesce well under X-rays, so grease tables have been used to recover such stones. Since diamonds are hydrophobic (meaning

A diagram of a kimberlite pipe with an underground mine and open pit at the top. *Illustration © Petra Diamonds*

they repel water), they stick to the grease while the rest of the wet concentrate runs off.

Today, X-ray transmission (XRT) sorting is replacing dense media separation, grease tables and X-ray luminescence because it is more effective at preventing the breakage of large diamonds during processing. Since XRT detects and separates minerals based on their specific atomic density, it can detect diamonds even if they are coated and have no fluorescence. The Lesedi La Rona and Sewelô diamonds (see page 63) were recovered at the Karowe mine in Botswana using XRT technology.

Underground mining initially requires a considerable investment in infrastructure and development before production can begin. However, once the shafts and tunnels have been created, it is an economical mining method that is capable of being automated.

A rock breaker at the Cullinan mine in South Africa. *Photo © Petra Diamonds*

Loaders at the Finsch mine in South Africa collect ore for tipping before it is delivered to the crusher. *Photo © Petra Diamonds*

Trucks transport ore underground at the Finsch mine. *Photo © Petra Diamonds*

A ground handling worker at the Finsch mine performs a routine inspection of a conveyor belt to ensure that production delays are kept to a minimum. *Photo © Petra Diamonds*

Gravel is run through a dense media separation washing screen at the Williamson plant in Tanzania. *Photo © Petra Diamonds*

A grease table at the Williamson plant is used for second recovery. *Photo © Petra Diamonds*

A sorter at the Finsch recovery facility in South Africa uses specialist equipment to ensure that the rough diamonds are categorized correctly and in time for collection. *Photo © Petra Diamonds*

## Diamond Sorting

After the rough diamonds are separated from the ore, they are delivered to sorting experts, who categorize them and assign them a value. Diamonds are sorted into parcels according to their shape, size, color and clarity, but within these categories there are thousands of variants that can affect the price. This is when gem-quality diamonds are separated from small, low-quality industrial diamonds, which are also called bort. Industrial-quality diamonds are used as drill bits and for cutting equipment. In recent years, they have even appeared in designer jewelry.

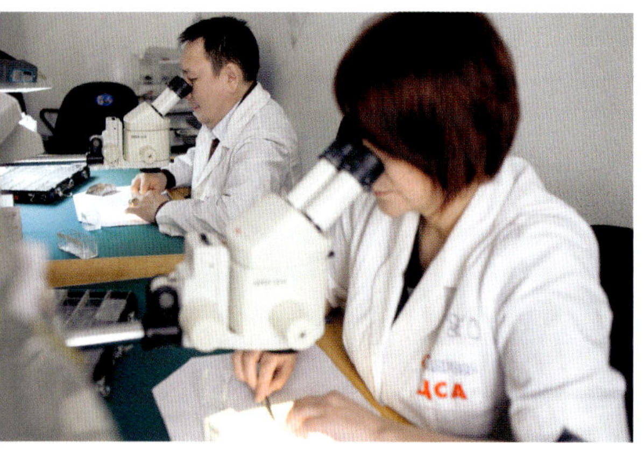

Workers analyze each crystal in transmitted light with a microscope to detect surface and internal defects. *Photo © Alrosa*

After rough diamonds are sorted, they are distributed to cutting centers and companies in various ways. Some diamond mining companies have their own cutting factories. Rough diamonds may also be sold to cutting facilities within the country where the diamonds are mined. De Beers sells its rough diamonds at events called "sights" to sightholders, which are a limited number of diamond companies with rights to buy diamonds in bulk directly from De Beers. Diamond rough is also sold at diamond bourses in cities such as Antwerp, Tel Aviv, Hong Kong, New York and Mumbai. Large, important rough diamonds may be sold at auction or sold privately to buyers.

The next chapter explains how the sorted rough diamonds are cut and polished after cutting facilities receive them.

A close-up view of rough diamonds being sorted. *Photo © Alrosa*

The rough diamonds are weighed during the sorting process. *Photo © Alrosa*

St Top
St Spt
St Spec
Cleav Top
Cleav Spt
Cleav Spec
Mac Spt
2 Blk Z
3 Blk Z
4 Blk Z
Fl Spt
2 Blk Cliv
3 Blk Cliv

A display of sorted rough diamonds.
*Photo © Alrosa*

# 4 The Evolution of Diamond Cutting

Nobody knows who first tried to shape a diamond or when or where they tried it, but according to diamond jewelry historian Jack Ogden, the dawn of diamond polishing and cutting in Europe had likely occurred by 1300, though evidence is sparse. One might assume that since India is the oldest source of diamonds, it is also where the art of diamond cutting began. However, the ancient Hindus believed that altering diamonds in any way would rob them of their magical powers, so they were not motivated to find ways to change their shape. As a result, well-formed transparent crystals with few flaws were more valued in India than distorted shapes.

Around the mid-1300s, European and Indian gem cutters began to repair diamond crystals that had rough exteriors or damaged edges or points. By that time, caravans were transporting rough from India to Venice, Italy, an established trade center. The first cutting style was a basic modification of the most common diamond rough shape, the octahedron, which resembles two pyramids stuck together base to base. Diamond cutting in the 1500s saw an evolution from the mere polishing or repairing of natural crystal faces to cutting the gems into new shapes. Cutters experimented with variations of rose cuts and cleavages, as they added facets. They also produced an array of new cuts by adding extra crown and pavilion facets to early table cuts, which were created by making a flat cut at the point of the octahedron.

Before discussing how diamond cutting evolved from simply modifying rough diamonds to transforming them into modern brilliant cuts, it would be helpful to first understand the crystal structure of diamonds and the processes involved in cutting current diamond styles.

## CRYSTAL STRUCTURE AND CUTTING

Crystal structure affects diamond in ways that are important to cutters. A diamond's cubic crystal system has four directions that are the weakest points in its molecular arrangement. In these directions, there are fewer carbon atoms and more space between them, so a well-placed blow can cause a flat break parallel to one of these directions. These are called cleavage directions.

Diamonds are almost uncompromising in the directions where the atoms are bonded tightly together. These directions are harder and, consequently, unsuitable for sawing and polishing.

Cutters use cleavage to divide distorted diamond crystals into more workable shapes and to remove major inclusions, thus improving the clarity and, therefore, the value of the finished gems.

Some crystals consist of two or more individual parts with common atomic planes but different orientations. These are called twinned crystals. The most common twinned diamond crystal is the macle, which we will discuss a bit later.

An important value factor for cuttable rough diamond is shape because it determines the ultimate yield. Shape is so important that diamond cutters have coined special terms to describe the rough's potential.

A sawable is diamond rough that will yield more weight if it is cut to produce two stones. An octahedron is considered a sawable and is the most expensive shape. Two round brilliants or princess cuts can be produced from it.

A makeable is irregular-shaped diamond rough that can be polished without sawing, cleaving or splitting. The final shape is a single stone that is often similar to the shape of the rough, and the yield is less than that of a sawable.

A splittable (also known as a cleavage) is diamond rough that can be divided into small but valuable segments by lasering or cleaving. These include diamonds that were imperfectly formed as well as those that are fractured or were split along a cleavage place at some stage of their existence. According to Eric Bruton in his book *Diamonds* (1978), "*Cleavages* can contain very large stones of the very highest quality. For example, the Cullinan would have been graded as a *cleavage* had

Diamond octahedron (10.82 carats).
*Diamond courtesy of Pala International; photo by Jason Stephenson*

## How Diamond Crystal Shape Affects Value

More Valuable

**sawable**
round brilliants or princess cuts

**makeable**
round brilliants

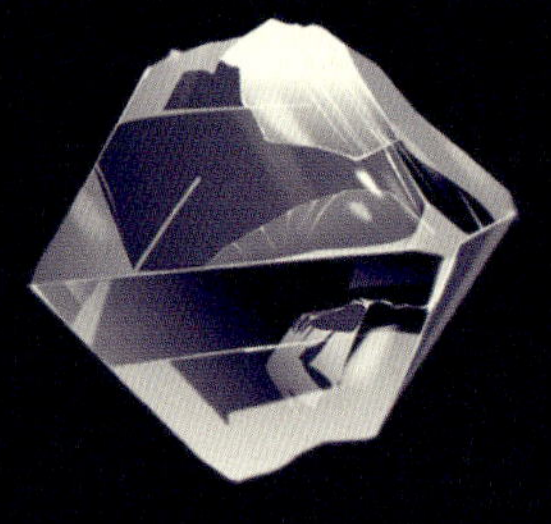

**splittable**
round brilliants or fancy cuts

**macle**
fancy cuts

**flat**
baguettes or similar small stones

Less Valuable

The shape of the diamond rough determines the size and shape of the stones fashioned from it. Sawables and makeables can yield larger gems and more profit than macles and flats of the same weight. *Image by Peter Johnston © GIA; reprinted with permission*

it reached the sorting rooms." Cleavages range in form from near-sawables or near-makeables with distorted or broken faces to irregular lumps of diamond with no recognizable crystallographic features.

A macle, the fourth rough-sorting category, is a twinned crystal that has two opposing parts each with the same crystal shape. The two parts are oriented 60 or 180 degrees from each other, so the macle looks like a flattened triangle. Besides having opposing crystal directions, macles are also flat, which makes cutting very difficult. They also usually have a seam, called the twinning plane. Macles are often too shallow to yield round brilliants without significant weight loss, so they are usually used for fancy shapes like pears, triangles and hearts.

A flat is the lowest-priced rough category and is limited in its potential shape. It includes tabular crystals that are too thin for making brilliants.

Cutters must consider both marketability and weight retention when deciding the shape of the finished diamond. For example, even if well-shaped rough can be fashioned into one large round brilliant, it might be better to cut smaller diamonds from it because the smaller gems could be easier to sell than a single large one. On the other hand, round brilliants are typically the most marketable choice, but they might not be the most profitable shape to cut depending on the diamond rough. The rough can potentially lose too much weight if it is always cut as a round.

After the rough has been sorted according to shape, it is sorted for clarity and color. Clarity can affect the yield of diamond rough if inclusions do not allow the cutter to orient the rough for the best weight retention and yield. Evaluating color in rough is different than judging it in faceted diamonds because surface texture or stains can affect the color of the rough diamonds, and the color may not be evenly distributed throughout the crystal. Sometimes rough is sorted by color before it is sorted by shape, especially in the case of fancy color diamonds.

A 3.32-carat colorless macle displaying superb luster, transparency and clarity. Even though macles sell for less than octahedrons of the same size and quality, macles like this one fetch high prices and are coveted. Specimens of this collectible weight and quality are increasingly uncommon and generally are older specimens from collections. *Diamond macle and photo courtesy of The Arkenstone*

A blade mechanically saws a rough diamond. *Photo © Asian Star Group*

The optional second stage involves splitting or sawing the rough diamond to create a more attractive shape, remove distracting inclusions or produce fragments for use in tools. Cleaving is the act of dividing a diamond in two along the grain with a blow of a hammer on a cleaving knife. Sawing is done in a non-cleavage direction by cutting through the diamond with a blade charged with diamond dust.

A traditional polishing wheel. *Photo © Asian Star Group*

The introduction of the rotary diamond saw around 1900 allowed cutters to create multiple stones from one piece of rough, instead of having to grind down the rough into one stone. It also helped retain weight from the original rough. In the late 1970s laser sawing, which substitutes a laser beam for the metal saw blade, was introduced. The laser burns into the rough and vaporizes it as it slices a narrow channel into the rough. Laser sawing is faster and more versatile than mechanical sawing and is the preferred method for dividing diamond rough today.

After cleaving, the diamond is rubbed against another diamond to grind it closer to its intended final form and create the girdle outline (the narrow rim around the stone). This third stage is called bruting or girdling. In the late 1980s automatic bruting machines that needed little supervision were introduced. One skilled cutter can monitor and adjust several machines at once, and two stones can be bruted simultaneously — each stone shaping the other in a more consistent roundness. Laser bruting was introduced in 1992 and is used mostly for fancy cuts because it can create shapes that are precise and symmetrical, such a hearts, ovals, marquises, pears and so on.

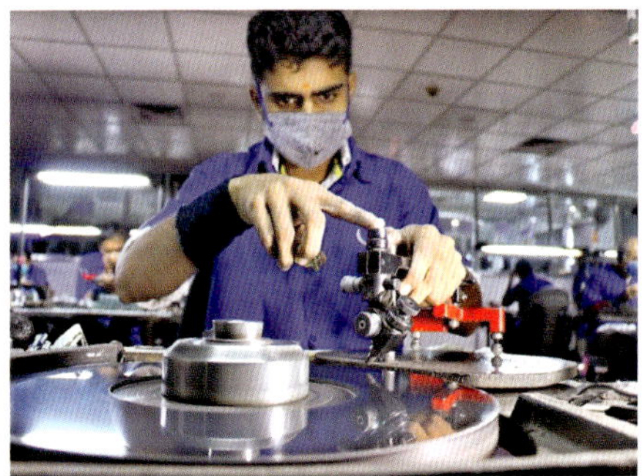

A worker polishes at a jumbo wheel. *Photo © Asian Star Group*

The fourth stage is called cross-working (blocking), during which a worker called a blocker places the first 17 or 18 facets: the table and

Workers oversee automatic polishing. *Photo © Dharmanandan Diamonds Pvt. Ltd.*

A worker facets the crown and pavilion on a polishing wheel. More than 95 percent of the total production in the Dharmanandan facility is done by hand. *Photo © Dharmanandan Diamonds Pvt. Ltd.*

usually eight facets on the crown, eight on the pavilion and sometimes a culet. The diamond is held against a rotating iron wheel or spinning horizontal disc called a scaife, which is charged with diamond powder.

Blocking must be done carefully because it establishes the diamond's basic symmetry. If the blocked diamond is uneven or if the crown and pavilion facets are not aligned at the girdle, it must be blocked again, resulting in weight loss.

The final 40 facets of a brilliant cut are polished during a process called brillianteering. GIA and many diamond dealers collectively refer to the blocking and brillianteering processes as the polishing phase.

## MACHINE ADVANCEMENTS IN CUTTING

An illustration of workers polishing diamonds by hand in a 19th-century diamond factory. *Historical Images Archive/Alamy Stock Photo*

When the scaife was first introduced in the 15th century, workers pushed and pulled a lever to spin a drive wheel that spun the steel or iron scaife.

Water power (using a mill wheel) was introduced in the 1530s. Still, many scaifes used manual labor for turning the wheel. By the early 1800s several horse-powered scaifes appeared in Amsterdam, and by 1840 steam power had been introduced. Shortly afterward large-scale cutting factories, where central steam engines drove a group of scaifes, started springing up. Diamonds from Brazil discovered in the early 1700s were a source of stones for these factories and contributed to a surge in building larger diamond-cutting factories. As the supply of diamonds dwindled in Brazil, diamonds were discovered in South Africa.

In the mid-1890s, the first mechanical faceting dop was patented. It held the diamond firmly between steel jaws and made it easy for cutters to change a stone's position during faceting. By 1900 electric power had come to urban areas and was used to run the machines needed for sawing, bruting and polishing.

The invention of the rotary saw in the early 20th century played a major role in the development of the modern round brilliant. It allowed cutters to fashion an octahedron into two diamonds with larger tables and shallower crowns while minimizing weight loss from the rough, since the smaller top part of an octahedron could be turned into a finished diamond instead of being ground away.

Traditional diamond cutters would determine the best polishing directions for orienting the diamond for cutting. Sometimes that involved attempting to polish a facet and then stopping to check to see if the polishing was successful. This changed with the development of automated polishing machines like the Piermatic Diamond Polishing Machine developed by De Beers in the early 1970s. It enabled an operator to polish several diamonds at one time on a single machine as well as monitor multiple machines.

Today diamond cutting is not only mechanized, it is also computerized. Imaging software using a low-level laser can read the rough's entire surface to help cutters determine how to cut a diamond with the most weight and best quality from the rough. Modern diamond cutting technology has increased efficiency and brought standards of precision and symmetry that were difficult to achieve with conventional cutting methods. Nevertheless, the human eye is still required to verify that the cutting and polishing are done properly and to check diamond clarity. Automatic polishing is often reserved for small and lower-quality diamonds. Larger high-quality diamonds may still be polished by hand using computer-aided technology to determine how to cut and polish the diamonds.

For example, Dharmanandan Diamonds, whose polished diamonds regularly receive "triple excellent" grades from GIA, cut and polish more than 95 percent of their diamonds by hand, though all of their diamonds are planned using computers. There is no size limit on automatic polishing machines, but Dharmanandan Diamonds only uses them for half-carat and smaller diamonds of lower color and clarity. As for their production of manually polished diamonds, the size ranges from 0.01 carats to 15 carats. Diamond polishing by many other companies is fully automated.

A worker examines the cut and clarity of a newly polished diamond through a quality control microscope. *Photo © Dharmanandan Diamonds Pvt. Ltd.*

Workers perform quality control on polished diamonds. *Photo © Dharmanandan Diamonds Pvt. Ltd.*

## The Development of Diamond Cuts

The first cutting styles were simple modifications of typical diamond rough shapes, such as the octahedron. The surfaces were smoothed by rubbing the diamonds on boards coated with diamond grit and olive oil. As technological advancements gave cutters more power and flexibility to control diamond cutting, the cuts became more sophisticated and the techniques more refined, resulting in the sparkle and magic that we attribute to diamonds today.

**Parts of a Cut Diamond**

Table

CROWN VIEW

Girdle

Crown

Pavilion

SIDE VIEW

Culet

PAVILION VIEW

### POINT CUT

The point cut was the earliest diamond cutting style and was popular from the 1300s into the 1500s, though the style has reappeared in contemporary jewelry. The stones were polished with diamond grit and olive oil on a stationary polishing surface. The faces of a diamond octahedron crystal are almost impossible to polish

Point cut of a sawn octahedron in a ring by artist-jeweler Todd Reed. *Photo by Azad*

A yellow diamond octahedron from the Lulo mine in Angola. *Crystal and photo © Lucapa Diamond Company Limited*

Transparent octahedral diamond crystal. *Photo © Paul Cassarino of the Gem Lab*

because of their orientation to the atomic structure of the diamond, so the cutters had to cut at a substantially different angle to smooth the faces. That is why point-cut diamonds have angles below or above those of natural octahedrons. Early true point cuts are recuts of diamond octahedrons that had imperfections or damage on the surface and were polished to create a good double pyramid shape. Point cuts will have sharper edges and different angles from natural octahedrons. Natural octahedrons are not considered point cuts.

### TABLE CUT

The table cut is an octahedron with its top point flattened to a square facet called a table. Often the lower point of the stone was also removed to form a smaller square or rectangular facet called a culet. This gave the stone a total of 10 facets: five on the crown (top) and five on the pavilion (bottom). When such stones are viewed from above, the table cut looks like a box within a box. The table cut greatly improved the amount of light returned to the viewer, making diamonds brighter than point-cut gems. When the sharp point at the top of a diamond was damaged through normal wear, the point-cut was reshaped by cutting a flat "table" on the top, creating the more attractive table cut. Table and point cuts dominated diamond jewelry through the 1500s and 1600s, but they are also found in jewelry of the 1700s. Variations included rectangular tables that were cleavages from diamond crystals.

A close-up view of the upper and lower parts of an 18-karat gold and diamond cross pendant necklace, circa 1730, set with table cuts, rose cuts and senailles (small diamond chips with a few facets). *Pendant and photos from Adin Fine Antique Jewellery (AntiqueJewel.com)*

The front and back of a Georgian-era pendant cross set with table-cut and rose-cut diamonds (circa 1750). The rose cuts are set on foil, which was a common technique used in the 1700s to improve the appearance of diamonds. The mounting is made of 18-karat rose gold, and the piece is believed to be of French or Belgian origin. *Pendant and photos from Adin Fine Antique Jewellery (AntiqueJewel.com)*

An Art Deco portrait diamond ring with emerald accents and a round brilliant diamond border. *Ring from GeorgianJewelry.com; photo by Zachary Mial*

Some view the portrait cut as a variation of the table cut. Like other old cuts, portrait cuts can be found in both antique and contemporary jewelry. The portrait cut is an extremely thin diamond, and it originally had a single row of facets around the edge of a very large table, which resembles a thin sheet of glass. It was historically used to visually enhance and protect miniature paintings, acting as a window to the image set below the diamond. Modern designers have also used portrait-cut diamonds, usually with more than a single row of step-cut facets and no underlying photo or painting. The lack of brilliance this cut offers attracts those looking for understated elegance in their jewelry. Besides being the perfect way to present extraordinary diamond clarity, the portrait cut looks both ancient and modern simultaneously.

A contemporary portrait-cut diamond ring by Anup Jogani. *Photo courtesy of Anup Jogani*

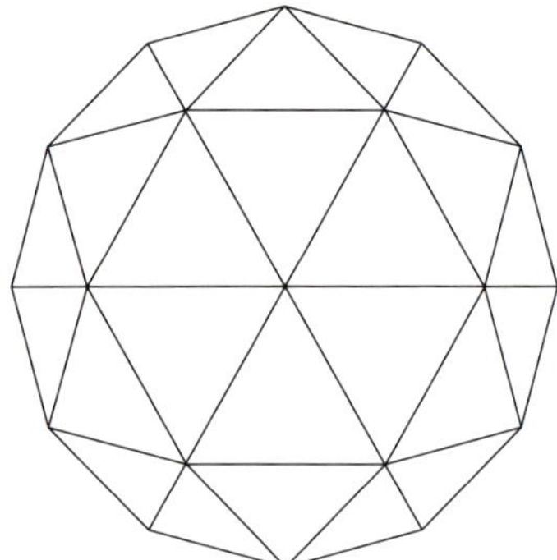

ROSE CUT

## ROSE CUT

Rose-cut diamonds are dome shaped with flat bottoms, and they have rose-petal-like triangular facets that usually radiate out from the center in multiples of six. From above, the rose cut may appear round, oval or pear shape. The shape of the original diamond crystal largely determines the shape into which it will be cut. Rose cuts were ideal for flatter and thinner rough, such as cleavages, broken crystals and macles.

In the 1400s rose-like cuts were in wide use and gradually evolved into the classical rose cut by the 1500s. Most were triangular and flat bottomed, and the top was divided into any number of facets. The name rose cut was derived from the fact that the cut diamond is meant to look like the opening of a rose bud.

Rose-cut diamonds have more sparkle than table-cut diamonds but not as much as brilliant cuts. During the 18th and 19th centuries both Amsterdam and Antwerp specialized in rose cuts.

Rose-cut diamonds foiled and close backed in a silver and yellow gold ring, circa 1700–1740. *Ring from GeorgianJewelry.com; photo by Zachary Mial*

A rose-cut greenish diamond accented by eight table-cut diamonds in a Georgian ring, circa 1750 or earlier. All the diamonds are set closed backed and are likely foiled beneath. The pointed metal edge work, often termed pie crust, is typical of a technique used in Europe during the early to mid-1700s. *Ring from GeorgianJewelry.com; photo by Zachary Mial*

Rose-cut Fancy yellow diamonds in a bracelet by Hubert Jewelry. *Photo by Diamond Graphics*

Rose-cut diamonds in a Georgian-style ring of silver over 14-karat gold. *LangAntiques.com; photo by Cole Bybee*

Rose-cut pear-shaped diamond set in a gold ring. *LangAntiques.com; photo by Cole Bybee*

An Art Deco briolette pendant necklace. *Photo © Adin Fine Antique Jewelry (AntiqueJewel.com)*

In recent years, rose cuts have become very popular again. In fact, most rose cuts on the market are new, typically cut in India or Turkey. The demand started with designers who use them for reproductions. The newer rose cuts tend to be more symmetrical, while the shape and facets of most old rose-cut diamonds are typically irregular.

A variation of the rose cut is the briolette, a teardrop-shaped stone with triangular or diamond-shaped facets all around it. An example is the 2.30-carat briolette shown on the left, which is suspended on an Art Deco platinum pendant circa 1920. It is set with graduated old European–cut and antique cushion-cut diamonds following the drop outline. This briolette is believed to be of Belgian origin and was exhibited in 2013 at the Antwerp Diamond Museum.

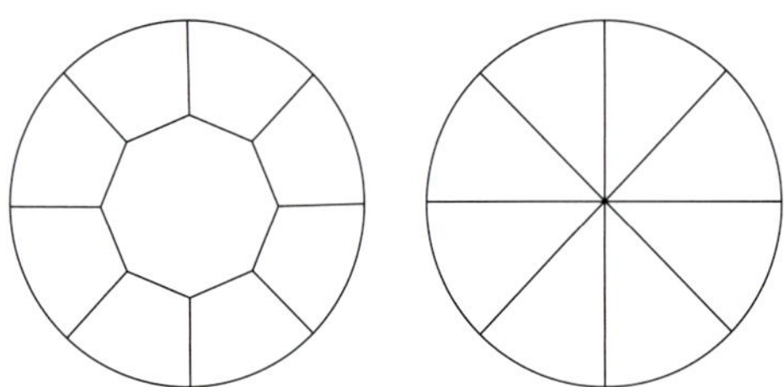

**SINGLE CUT**

## SINGLE CUT

After creating the table cut and rose cut, Europeans experimented with other cutting styles. In the mid-1600s the single cut, a round-shape cut from roundish rough, was introduced. It sparkled more than the table cut because there were more facets: a table, eight crown facets, eight pavilion facets and almost always a culet (the small facet on the pointed bottom of the pavilion). This cut served as the basis for the modern brilliant cut, and it is still used for small diamonds, most commonly weighing less than one-tenth of a carat. The advantage of using single cuts is that they cost less than modern round brilliants of the same quality (which have 57 to 58 facets) because less labor is involved in cutting a diamond with only 17 or 18 facets. In addition, if you are not looking at small diamonds under magnification, you will not notice much of a difference in brightness and sparkle between a single cut and a round brilliant.

Single-cut diamonds in a black onyx and 18-karat gold ring. *Photo © Heritage Auctions (HA.com)*

## DOUBLE CUT (MAZARIN CUT)

The double cut, or Mazarin cut, was also introduced in the 1600s. It is an octagonal shape with 17 facets above the girdle and 17 below, including the culet. The Mazarin cut was named after France's Cardinal Jules Mazarin (1602–1661), who is said to have been the first to have a diamond cut in this style.

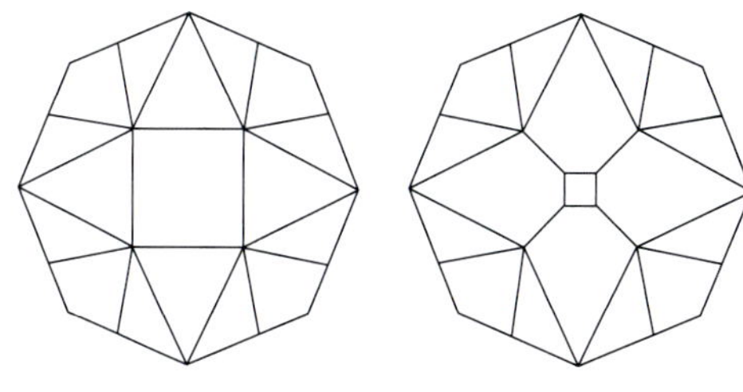

**DOUBLE CUT**

## TRIPLE CUT (BRILLIANT CUT, OLD MINE CUT)

New developments in diamond cutting were aided by the industrial revolution and the discovery of diamonds in Brazil in the 1700s. The greater availability of diamonds resulted in lower prices. Suddenly even the upper-middle class could afford diamonds, and to be able to own diamonds like the rich was exciting. By the mid-1700s this new popularity of cut diamonds and the abundance of rough encouraged diamond cutters to experiment with new faceting styles.

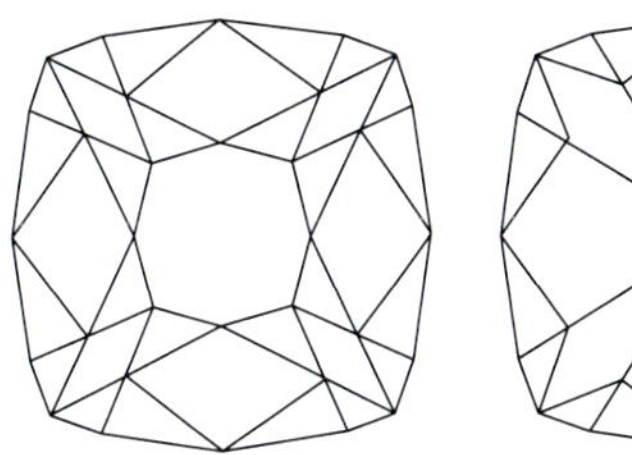

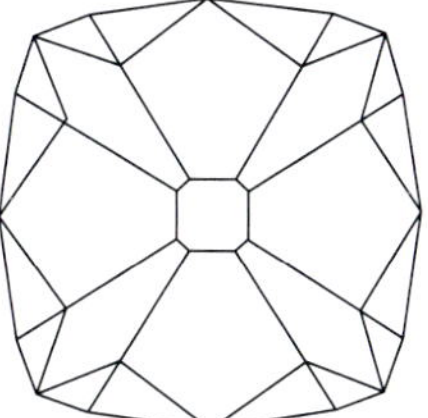

**TRIPLE CUT**

One of those was the triple cut, which was simply called "the brilliant" by David Jeffries, a diamond dealer and author of *A Treatise on Diamonds and Pearls* (1751). Today antique dealers call it the old mine cut. This cut is a cushion shape (a rectangular or squarish shape with curved sides and rounded corners) with 58 facets, similar to the modern brilliant but with a higher crown, a greater total depth, a smaller table and a larger culet. The girdle can often be inconsistent and very thin in places, and the lower half facets on the pavilion are shorter, which creates a different pattern of bright and dark areas and larger flashes of spectral colors than those found on a round brilliant–cut diamond.

Historically, the term old mine cut originated from the fact that most diamonds with this cut had been mined in India or Brazil instead of in the newer mines of South Africa, which were established starting in the 1870s. Today old mine cut refers to the cutting style of any 58-facet brilliant-cut diamond with short lower half facets, a large culet, a small table and a cushion or squarish outline. It is also simply called a mine cut. From the early 1700s to the late 1800s, the old mine cut was a common diamond cut, and it characterizes Georgian era (1714–1837) and Victorian era (1837–1901) jewelry.

The faceup and side view of an old mine-cut diamond. Note the high crown, thin girdle and irregular shape. *Photos © Renée Newman*

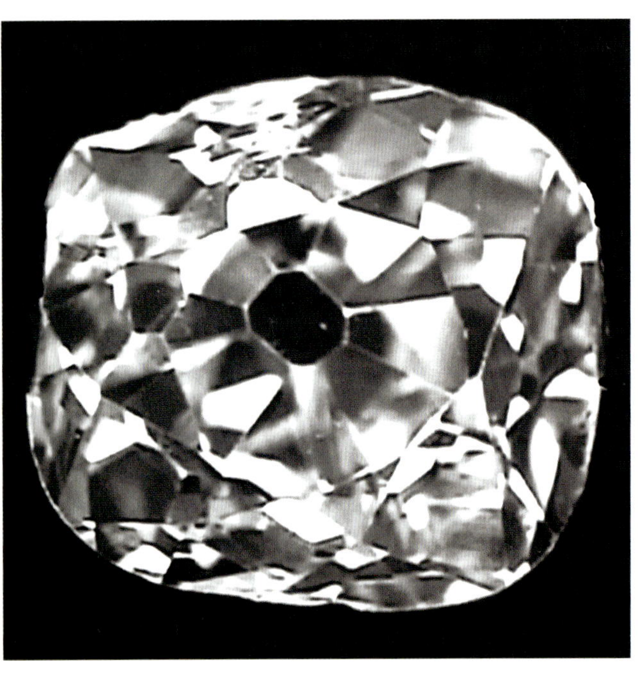

A 1.13-carat old mine-cut diamond recut into a 0.78-carat radiant cut. *Photos © Paul Cassarino of the Gem Lab*

A 1.71-carat old mine-cut diamond recut into a 1.39-carat modern round brilliant cut. *Photos © Paul Cassarino of the Gem Lab*

Old mine cuts are sometimes recut as modern cuts to make them brighter and more symmetrical. In other instances, antique cuts are only repolished to remove scratches and chips and increase luster.

The top and side view of a 7.51-carat old mine-cut diamond that was repolished. *Ring and photos courtesy of Abe Mor Diamond Cutters*

Old mine-cut and cushion-cut diamonds are set in silver topped with yellow gold in this English Georgian-era brooch, circa 1840. *Brooch from GeorgianJewelry.com; photo by Zachary Mial*

A Victorian old mine-cut diamond ring accented by enamel applied on 18-karat gold. *Photo © Heritage Auctions (HA.com)*

Old European-cut diamond in a ring.
*LangAntiques.com; photo by Cole Bybee*

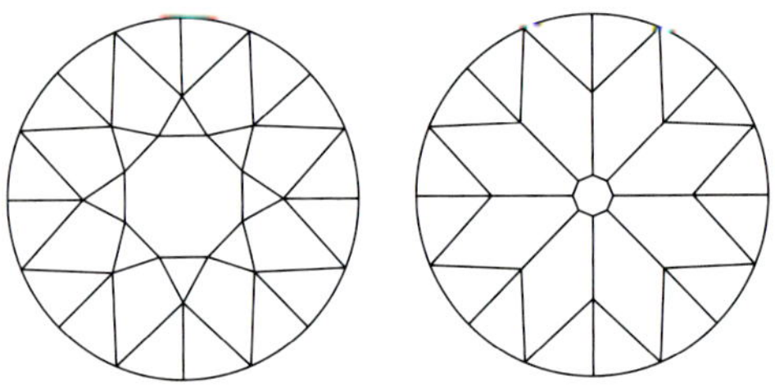

**OLD EUROPEAN CUT**

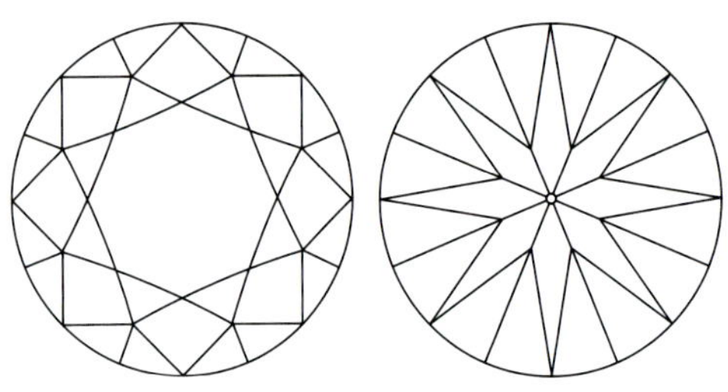

**MODERN ROUND BRILLIANT CUT**

## OLD EUROPEAN CUT

The old European cut usually has a small table, a high crown, a large culet and short lower half facets (a.k.a. lower girdle facets). It is similar to the old mine cut, except the old European cut is round and sometimes less bulky. It is the direct ancestor of the modern round brilliant cut and is the most common diamond cut seen in antique jewelry. Even though both old European and modern round brilliant cuts are round and typically have 58 facets, they have different patterns of bright and dark areas when viewed faceup. The old European cut has been described as having a chunky pattern of light that enhances a diamond's fire by providing broader flashes of color when the gem is viewed in spotlighting, whereas a modern round brilliant has narrower pavilion facets, which provide a splintery pattern that makes the diamond look brighter and more sparkly. The old European cut's wider pavilion facets and shorter lower half facets and the modern round brilliant's thinner pavilion facets and longer lower half facets are compared in the diagrams on the left.

According to GIA.edu, GIA identifies round diamonds as classic old European cuts using four criteria:

1. **Table size:** less than or equal to 53 percent of the average girdle diameter;
2. **Crown angle:** greater or equal to 40 degrees;
3. **Lower half facet length:** less than or equal to 60 percent of the total distance between the girdle and the culet;
4. **Culet size:** slightly large or larger.

If a diamond meets three out of the four criteria, it will still receive an "old European cut" designation on a GIA grading report. A report for older cuts of diamond simply provides the gem's measurements and grades for color and clarity, as GIA only issues cut grades for modern round brilliants. GIA says these parameters were derived from historical definitions of the old European cutting style, staff observations and discussions with trade professionals. Simply stated, older-style diamonds should not be judged by standards that they were never fashioned to meet. If they were, they would likely receive a "Fair" or "Poor" cut grade. Dealers marketing the unique appearance of these diamonds also contend that applying modern cut-grade criteria would unduly penalize their stones.

Round faceted diamonds were somewhat uncommon before the 19th century, but they became popular in the late 1800s because of advances in cutting technology that helped improve diamond appearance.

The 1874 invention of the bruting machine for rounding a diamond mechanized that part of the cutting process, giving cutters much greater precision and control. Diamond cutters were able to give diamonds a more perfect round shape, and the public responded. By 1900 it was the most popular shape for diamonds.

Old European cuts are found in jewelry from the late Victorian, Edwardian and Art Deco eras. Designers today are also setting them in modern and vintage-style jewelry. True old European cuts with high color grades are extremely rare, as most high-color-graded older-cut diamonds have been recut to modern shapes.

According to Debra Sawatzky, an antique jewelry specialist, there has been an unusually high demand for old European cuts. It is sometimes easiest to meet this demand by recutting round brilliants, so buyers should not assume that rings mounted with old European cuts are antiques. Furthermore, even if the diamonds are truly old, this does not necessarily mean the ring is old. It is common for jewelers to remount old stones in new jewelry.

Crown and pavilion views of an old European-cut diamond. *Photos © Paul Cassarino of the Gem Lab*

Old European-cut diamonds set in an 18-karat gold Victorian brooch. *Photo © Heritage Auctions (HA.com)*

Old European-cut diamonds and platinum Edwardian ring set. *Photo © Heritage Auctions (HA.com)*

## MODERN ROUND BRILLIANT CUT

A Boston diamond cutter named Henry D. Morse is credited with proposing the proportions and angles for the modern round brilliant cut in the late 1800s. He believed that diamonds should be scientifically cut for beauty and not weight retention. Most other cutters at the time cut whatever shape the diamond would yield, with little regard to the angles they used. Morse experimented with different angles and changed the crown and pavilion angles so that they would not be so steep. This resulted in brighter diamonds with more sparkle. Nevertheless, the old European and old mine cuts remained popular into the 20th century.

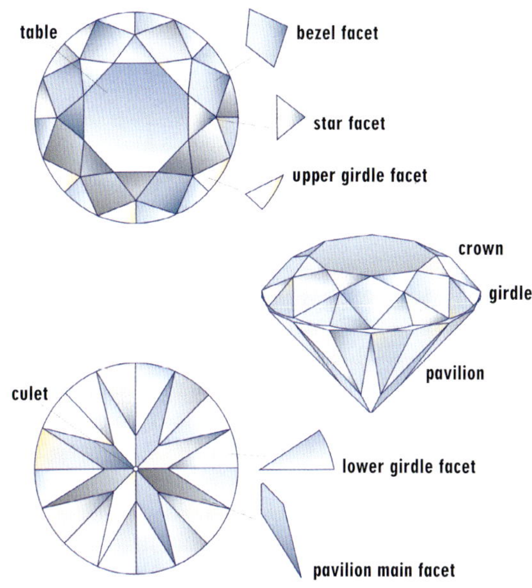


A diagram of the modern round brilliant cut. *Illustration by Peter Johnston © GIA; reprinted with permission*

Morse was also the first to introduce the circular diamond. Author Al Gilbertson notes in *American Cut: The First 100 Years*, "Although his revolutionary methods for bruting led to rounder brilliants, Morse mostly cut cushion-shaped or squarish diamonds in his shop. The round shape was too new to be in high demand."

The modern round brilliant cut has a lower crown, a smaller culet and a longer lower half facet length than its predecessors, but it has the same number of facets — 33 on the crown and 25 on the pavilion (including the culet, which is optional). Well-cut round brilliants have a pavilion angle at or about 41 degrees. Since about 1920 it has been the best-selling diamond cut. In 1919 diamond cutter Marcel Tolkowsky helped popularize the modern round brilliant by publishing his recommendations for its best proportions in a book called *Diamond Design*.

Today the girdles of high-quality brilliant-cut diamonds are usually faceted to give the diamond a more finished look. Bruted, unpolished girdles have a rough texture where dirt can collect and eventually affect the perceived color of the diamond. However, these facets do not influence the light performance of the diamond.

The old European cut did not immediately develop into the modern round brilliant cut. As cutters continued to experiment with proportions, there were many incremental steps between the two cutting styles. Consequently, some 58-facet round diamonds cannot be classified as either an old European cut or a modern round brilliant cut. Sellers of these diamonds call them transitional cuts. Dealer

Michael Goldstein says that these stones usually have the overall faceting of modern brilliants but still might have a slightly more open culet and shorter pavilion main facets.

According to Goldstein, the term transitional cut can also refer to a stone that is a cross between a cushion cut and an old European cut. A transition is the process of changing from one cutting style to another, hence the term transitional. (This term is often encountered by buyers shopping for vintage or antique jewelry.)

After GIA introduced its diamond cut grading system in 2005, it became apparent that expectations for transitional cuts are different than those for modern round brilliants. As a result, GIA decided to add a new description to address diamonds that fall outside the existing categories of both old European and modern round brilliant.

To establish judgment criteria and help define these diamonds, GIA met with members of the trade and borrowed many examples of transitional cuts from dealers. According to a 2013 article by Duncan Pay on GIA.edu, the goals were to

- create a new description for older-style rounds that reflects their historic cutting approach;
- prevent cut grades from being assigned to diamonds that were never fashioned to modern round brilliant parameters;
- prevent poorly cut modern round brilliants from passing through the GIA cut-grading system without receiving grades.

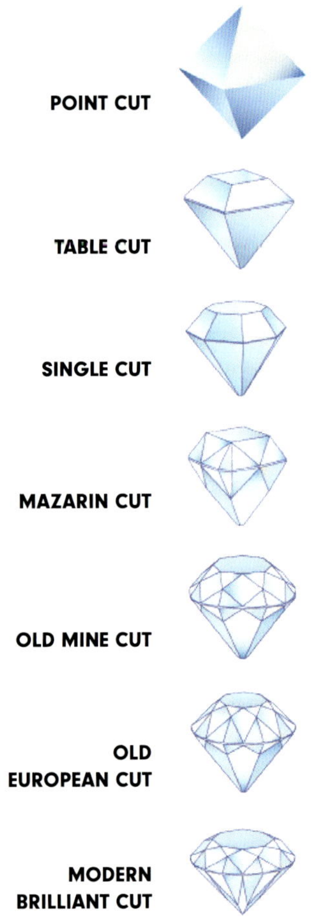


Diamond cuts evolved through many stages before the modern brilliant cut was developed. Each stage improved upon the one before it. *Illustration by Peter Johnston © GIA; reprinted with permission*

GIA agreed that a number of stones were not classic old European cuts or modern round brilliants. They also looked neither like well-cut nor poorly cut modern round brilliants. Afterward GIA decided to introduce a new term for 58-facet round brilliants — the circular brilliant. The requirements for a diamond to fall into this new category are as follows:

1. **Lower half facet length:** less than or equal to 65 percent of the total distance between the girdle and the culet;
2. **Star facet length:** less than or equal to 50 percent of the total distance between the girdle edge and table edge;
3. **Culet size:** medium or larger.

Crown and pavilion views of a circular brilliant diamond. *Photos © Paul Cassarino of the Gem Lab*

A Fancy gray-blue circular brilliant diamond flanked by two single-cut diamonds in an Edwardian platinum ring. *Photo © Heritage Auctions (HA.com)*

The circular brilliant designation acknowledges that the diamond is not a modern-day round brilliant, suggests a description for rounds of earlier times and keeps the historic old European cut definition unaltered. As mentioned earlier, GIA does not assign cut grades to older styles of cuts, such as the old European cut or circular brilliant.

The modern brilliant cut usually has 57 or 58 triangular-, kite- or lozenge-shaped facets that radiate outward around the stone and can be adapted to diamonds with shapes that are not round, such as the marquise, oval, pear, heart, cushion and trilliant, as shown on the right. Examples of fancy-shaped brilliant-cut diamonds are shown on the following pages.

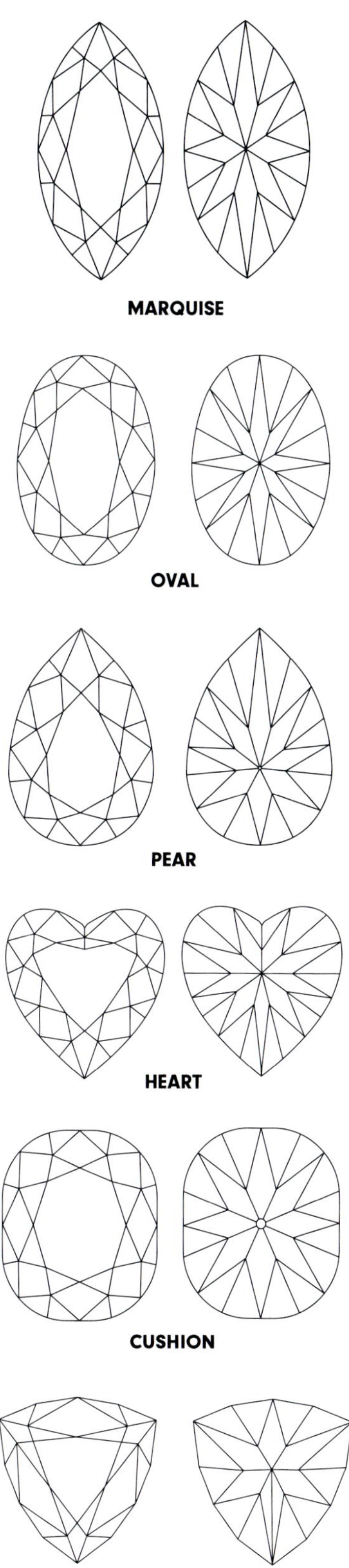

A 2.58-carat marquise-shaped brilliant-cut Fancy greenish-yellow diamond surrounded by colorless round brilliants. The band of this 18-karat gold ring is set with Fancy yellow round brilliants. *Ring and photo courtesy of Brian Denney of Gems of Note*

A 3.01-carat modified-pear-shaped brilliant-cut Fancy Light grayish-blue diamond accented by Fancy purplish-pink round brilliants set in platinum and rose gold. *Ring and photo courtesy of Brian Denney of Gems of Note*

An oval brilliant-cut Fancy Intense blue-green diamond surrounded by two pear-shaped Fancy Light pink diamonds and 18 Fancy Light pink round brilliants set in a platinum ring, accented with 18-karat rose gold. *Ring and photo courtesy of Brian Denney of Gems of Note*

A rectangular modified-brilliant-cut Fancy Deep pinkish-orange diamond surrounded by 10 rose-cut colorless round diamonds, two pear-shaped rose-cut colorless diamonds and an outer row of round brilliant diamonds. *Ring and photo courtesy of Brian Denney of Gems of Note*

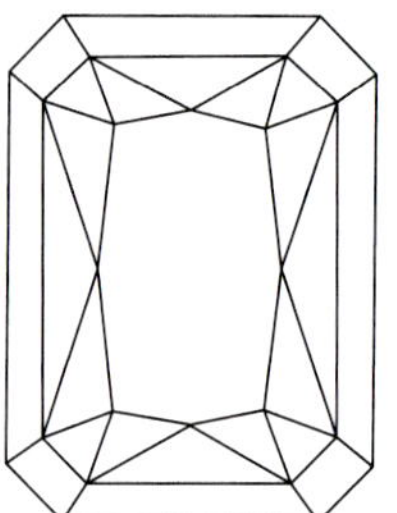

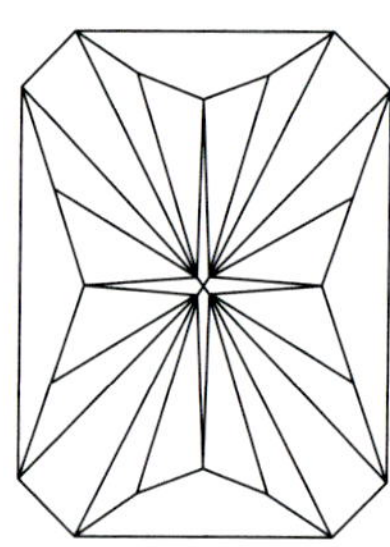

**RADIANT CUT**

## MIXED CUTS

In addition to brilliant-cut style facets, the mixed cut has four-sided and elongated facets parallel to the girdle, which are referred to as step-cut facets. The mixed cut is more often used on transparent colored gemstones rather than diamonds, but manufacturers also use the mixed cut to create unique faceting styles for diamonds.

Basil Watermeyer, a South African diamond cutter, is noted for creating the first combination of the brilliant-cut and step-cut styles for a diamond. He called this new cut the Barion cut. He applied for the patent in 1970 and received it four years later.

In 1977 Henry Grossbard of New York patented another mixed cutting style that he called the radiant cut. By the time the radiant cut patent expired, the diamond had become a well-accepted cutting style used by many manufacturers.

The center diamond of this ring is radiant cut. *Ring from J. Landau; photo by Leonard Derse*

The princess cut appeared around 1980. It is a square mixed cut that retains the 90-degree-angle corners and has brilliant-style facets on both the pavilion and crown and four elongated step cuts along the girdle of the crown. Also referred to as a modified brilliant, the princess cut can have 57 to 70 facets and variable proportions. Its pavilion facets widen toward the culet and narrow toward the corners to form a pattern resembling a four-pointed star. Los Angeles diamond cutter Barry Rogoff recommends placing a tiny facet on each corner of a princess cut to help prevent them from getting chipped. This process is called chamfering.

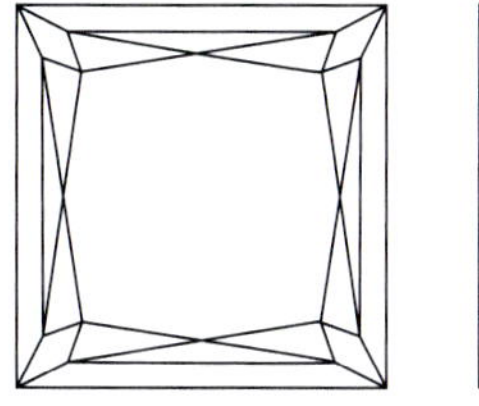

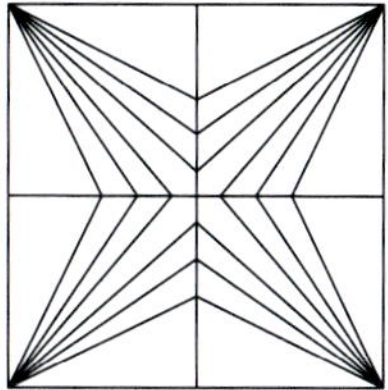

**PRINCESS CUT**

The center diamond of this ring is a princess cut. *Ring from J. Landau; photo by Leonard Derse*

Crown view of a princess-cut diamond. *Photo by Paul Cassarino of the Gem Lab*

The radiant cut and princess cut have inspired manufacturers everywhere to create a wide variety of patented diamond cuts, such as the examples shown below:

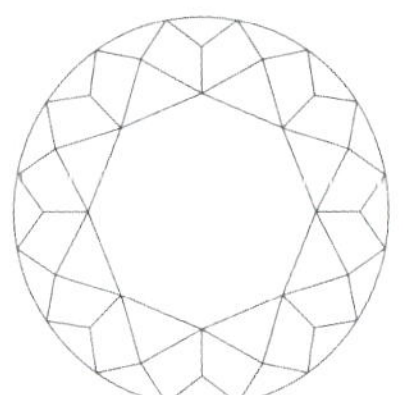
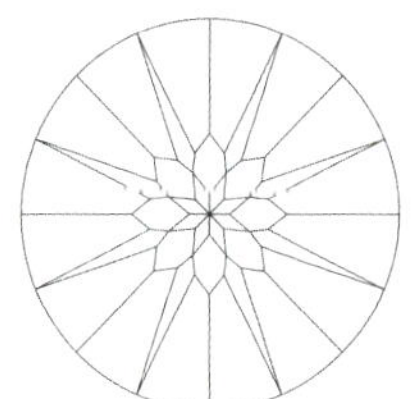
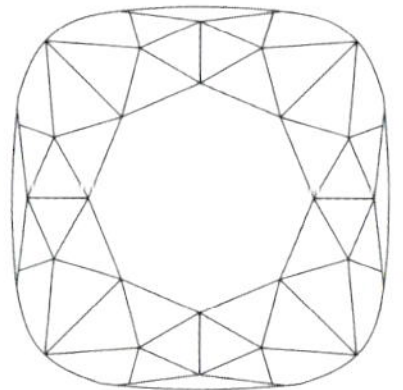
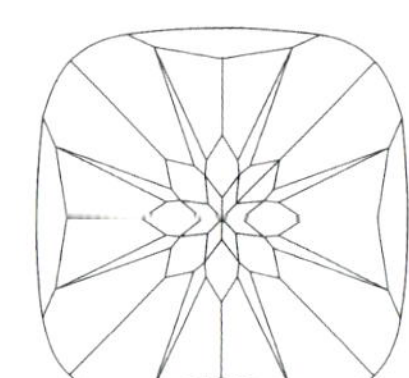

The Padma Cut Round and the Padma Cut Cushion by Dharmanandan Diamonds. *Photos © Dharmanandan Diamonds Pvt. Ltd.*

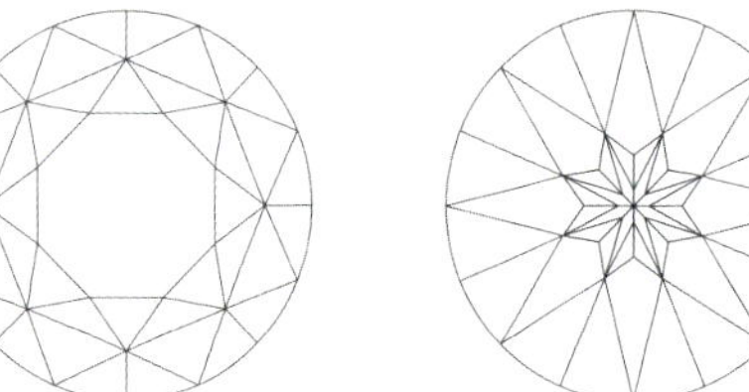
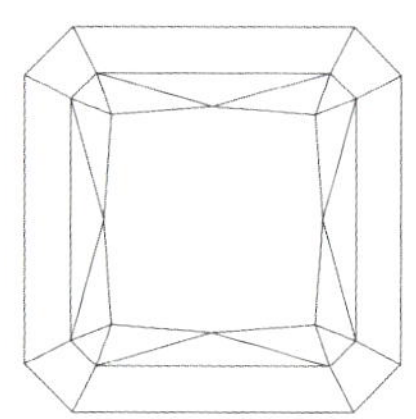
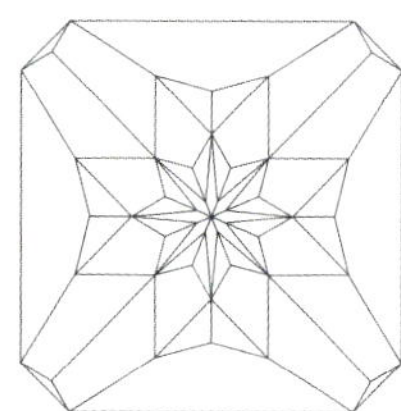

The Sirius Star 88 and Sirius Star Square were designed by Mike Botha and are manufactured under an exclusive global agreement by Dharmanandan Diamonds. *Photos © Dharmanandan Diamonds Pvt. Ltd.*

An emerald-cut Fancy Dark greenish-gray diamond flanked by two tapered baguettes set into an 18-karat white gold ring. *Ring and photo courtesy of Brian Denney of Gems of Note*

## STEP CUT AND EMERALD CUT

The table cut, which dominated diamond jewelry through the 1500s and into the beginning of the 1600s, is the precursor of the step cut. This cutting style has rows of facets that resemble the steps of a staircase when viewed from above and below. The facets are usually four sided, elongated and parallel to the girdle. Examples include the baguette and tapered baguette, styles that originated from narrow, table-cut diamonds called hogback diamonds.

If step cuts have beveled (clipped-off) corners, they are called emerald cuts because emeralds are often cut in this way. The bevels protect the corners and provide places where prongs can secure the stone. Emerald cuts are in essence step-cut octagonal rectangles. They tend to have more facets than baguettes. Emerald-cut diamonds are usually rectangular but are sometimes square.

In 1902 famed diamond cutter Joseph Asscher received a French patent for an emerald-cut style, which the company held until World War II. The cut had a higher crown with three rows of facets and a much deeper pavilion that has eight rows of facets and a large culet. Within only a few years, the Asscher company changed the shape to more closely resemble what is now understood as the Asscher cut, or simply an Asscher. The Asscher cut is typically square and has a slightly smaller table than Asscher's original design, but it still has a deeper pavilion than the emerald cut. Despite its greater depth, it is noted for its high brilliance and dispersion.

In 2001 the Royal Asscher Diamond Company introduced a new patented square emerald cut to the market called the Royal Asscher cut. It is an offspring of the original Asscher diamonds, but it has a total of 74 facets. Besides maximizing the brilliance of the stone, the new Royal Asscher uses the rough more efficiently. As a result, the diamond looks larger for its weight when viewed faceup, since the total depth is less.

**STEP-CUT BAGUETTE**

**EMERALD CUT**

Art Deco rings featuring step cuts, including an Asscher cut in the ring on the bottom left. *Photos © Heritage Auctions (HA.com)*

The diamond in the center of this ring is an Asscher cut. *Ring and photo courtesy of Abe Mor Diamond Cutters*

Today any generic square emerald cut that resembles the original patented Asscher is commonly called an Asscher cut in the jewelry trade.

The 1920s and 1930s saw a proliferation of straight-sided step cuts, which were ideal for the geometric Art Deco designs of that period. Baguettes continued to play a large role in jewelry design throughout the 1940s and 1950s. Step cuts in a variety of shapes remain popular today.

## SLICE CUT

A recent method of using rough with lots of inclusions and maximizing its faceup size is to slice the rough with a laser and create diamond slices ranging in thickness from about 1.5 to 2.5 millimeters. The surface of a slice may be entirely flat and smooth (synonymous with a portrait cut), or it might have some very low-angled large facets to create a sparkle effect from light reflected from its top surfaces. According to an article on the GIA.edu website, "The idea, say the manufacturers who create these pieces, is to cut the rough to show interesting patterns while keeping the stone's original outline. The more interesting the inclusions and their patterns, the higher the price."

The resulting cutting style, referred to as the slice cut, allows designers with an eye for nature's art to create one-of-a-kind designs. The slices are ideal for earrings because of their lower weight and price versus standard diamond cuts. They are also used for other types of jewelry, such as necklaces and rings.

The evolution of diamond cuts has gone full circle — from basic point cuts to table cuts, rose cuts, single cuts, double cuts, old mine cuts, old European cuts, modern brilliant cuts, unique patented cuts and now basic slice cuts and even uncut diamonds. However, a big difference now compared to 500 years ago is that all of these diamond cuts are available today, and they are all being used in contemporary jewelry. Some examples are shown on the next two pages.

Slice-cut diamonds in earrings and a ring by Hubert Jewelry. *Photos by Diamond Graphics*

Modern handcrafted rings by Single Stone featuring rose cuts, old European cuts and unique step cuts.
*Photos © Single Stone*

# 5

# The Evolution of Diamond Jewelry

The word "jewelry" is thought by some to be derived from the Latin and French words for "joy." Throughout history, a prime reason for wearing jewelry was to add a dimension of happiness and excitement to life. Prehistoric humans wanted more than just food, shelter and clothing, so they created jewelry from materials such as bone, teeth, rocks, shells, hair and feathers. Later, materials such as clay, glass, precious metals and gemstones were used.

## Early Diamond Jewelry

According to *Diamonds: An Early History of the King of Gems* by Jack Ogden, the earliest recorded diamond-set piece of jewelry is a 2,300-year-old gold ring with a pink sapphire flanked by two uncut diamonds. It was found in 1999 at Ai-Khanoum in what is modern-day Afghanistan. India was the source of those diamonds. Examples of Indian diamond-set jewelry date back to the first century BCE but are more abundant after the 11th century CE.

In the 1990s archaeologists excavated an ancient Roman necropolis in Vallerano, a municipality south of Rome. They found a gold ring set with a 0.15-carat octahedral diamond crystal in the tomb of a teenage girl, who they estimate was born around 150 CE. According to the Spring 2012 issue of *Gems & Gemology*, this ring is the only Roman diamond jewel with a known background. Most surviving Roman diamond rings date from the second half of the third century into the fourth, and these were generally set with one uncut diamond, though occasionally two.

A Roman gold and diamond ring from the third century CE. *© The Trustees of the British Museum*

A Venetian will from 1208 mentions a diamond ring, and written accounts of diamonds became more common as the 13th century progressed. Marco Polo (1254–1324), a Venetian merchant and explorer who traveled across Asia, wrote that Indian diamonds that reached Europe were only the rejects and that any diamond of importance was taken to Kublai Khan in China or other rulers.

In the mid-1300s diamonds became more abundant in Europe. King Richard II of England (1367–1400) had a diamond ring sent to the King of France as a diplomatic gift while trying to negotiate peace during the Hundred Years' War.

Even though the nobility lived in relative luxury during the 1300s, most Europeans endured filthy living conditions. The bubonic plague that struck Europe, Africa and Asia in the mid-1300s, known as the Black Death, killed one-third to one-half of the population, and peasants had the highest mortality rates.

The Black Death had profound effects on the course of history. With a sudden shortage of labor, the value of the peasantry increased. Many workers decided to start businesses and move from the countryside into cities, and a new social class of merchants emerged. Some historians theorize that the devastation of the plague experienced in Italy contributed to the cultural rebirth known as the Renaissance, which lasted from about 1400 to 1600. The Renaissance was a revival of Greek and Roman philosophy and art, and this period is especially recognized for its artistic developments and the contributions of Leonardo da Vinci (1452–1519) and Michelangelo Buonarroti (1475–1564). The invention of the printing press by Johannes Gutenberg in the mid-1400s was an important technological advancement that occurred during the Renaissance and led to the mass communication of ideas and higher literacy rates, further strengthening the emergence of the middle class.

**Carat Versus Karat**

The carat is a unit of mass equaling 200 milligrams (or 0.007 ounces). It was adopted as a standard measurement for diamonds, pearls and colored gems in 1907 at the French General Conference on Weights and Measures (Conférence générale des poids et mesures). However, it was only in July 1913 that it became a standard in the United States. Before then, the U.S. value of a carat was 205.3 milligrams.

In the U.K., the term "carat" is also a unit of measurement for gold purity. In America, the "c" in "carat" has been replaced with "k" to avoid confusion, resulting in the word "karat." One karat is 1/24 pure gold, so 24 karats is pure gold. Gold that is 18 karats in most countries is 18/24 parts, or 75 percent, pure gold. In Europe 18-karat gold is usually stamped "750," but in the U.S. an "18K" stamp is more common.

## Period Jewelry (European and American)

Jewelry historians and antique dealers often describe styles of jewelry with a period name based on the rule of a particular monarch or an art movement. Many of the periods overlap, and the beginning and starting dates vary depending on the historical source. The dates in this section are largely based on Christie Romero's *Warman's Jewelry*, Gail Levine's *Auction Market Resource* and Anna M. Miller's *The Buyer's Guide to Affordable Antique Jewelry* as well as information from dealers who sell antique and estate jewelry.

### Period Jewelry Terminology

**Antique jewelry:** Any jewelry that is 100 or more years old, as defined by the United States Customs Bureau. Merriam-Webster's dictionary defines the term antique more loosely: any work of art or the like from an earlier period.

**Heirloom, estate or vintage jewelry:** Previously owned jewelry that is typically passed on from one generation to another. It can range from a few decades to 100 or more years old. Vintage and estate jewelry can also refer to jewelry of an earlier era that has not been previously worn.

**Collectibles:** Items gathered from a specific designer, manufacturer or time period. The items are collected according to the buyer's interests. Normally these pieces are no longer in production, but they do not have to be as old as antiques. For example, Retro jewelry pieces are considered collectibles, but they are not yet true antiques – hence the phrase, "antiques and collectibles."

**Circa dating:** Establishing an approximate date of origin for a jewelry piece. Circa dating covers a 10-year window on either side of the date. A circa date of 1900 means the piece was probably made some time between 1890 and 1910.

Below is an outline of jewelry eras starting from the 18th century, followed by brief descriptions and some examples of pieces from each period.

| | |
|---|---|
| **Georgian** | 1714–1837 (reigns of King George I to King George IV) |
| **Early Victorian** | 1837–1860 (reign of Queen Victoria) |
| **Mid-Victorian** | 1860–1885 (reign of Queen Victoria) |
| **Late Victorian** | 1885–1901 (reign of Queen Victoria) |
| **Art Nouveau** | 1890–1914 |
| **Arts and Crafts** | 1890–1914 |
| **Edwardian** | 1890–1915 (reign of King Edward VII between 1901–1910) |
| **Art Deco** | 1915–1940 |
| **Retro** | 1939–1950 |
| **Mid-Century** | 1950–1970 |
| **Modern** | 1970–present |

## GEORGIAN (1714–1837)

Most diamond jewelry made during the Georgian period was hand-fabricated with 18- or 22-karat gold. In the early 1800s in England, gold filigree, an intricate form of metalwork, was popular. Diamonds were commonly set in silver over gold to bring out their whiteness. Diamond brilliance was intensified by foil backing the stones in closed-back mountings. The backs began to open up in the late Georgian period, after 1780.

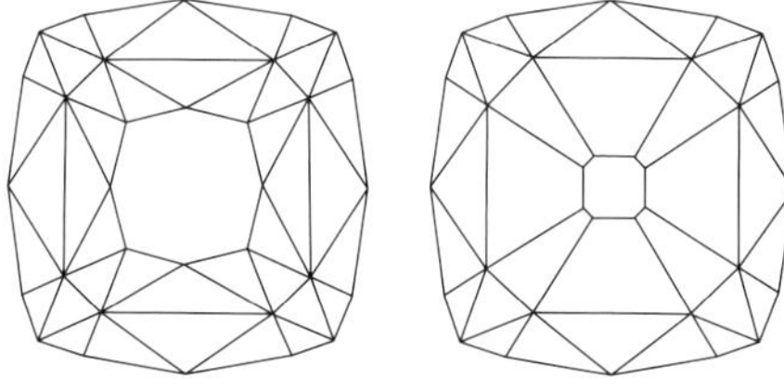

**BRAZILIAN CUT**

The discovery of diamonds in Brazil during this period along with the advancement of cutting techniques led to variations of the 58-facet cushion-shaped old mine brilliant cut. Two variations were the Brazilian cut and the Lisbon cut, shown on the right.

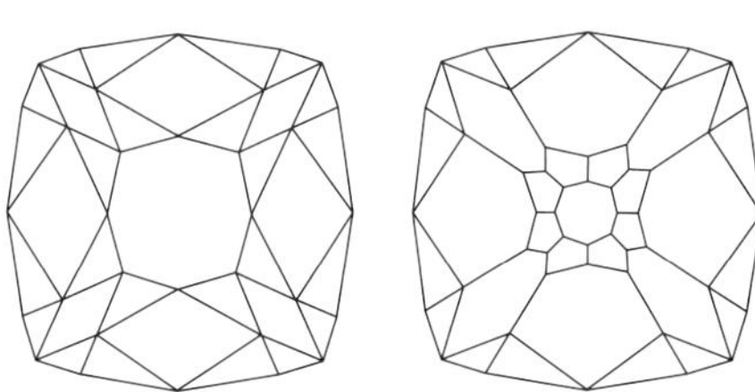

**LISBON CUT**

The themes of Georgian jewelry designs often came from nature: flowers, leaves, acorns, birds and feathers. The designs began to combine Brazilian diamonds with colored gemstones, such as aquamarine, garnets, green beryl and pink topaz.

A very rare English ring, circa 1760, with rose- and old mine-cut diamonds, foiled and set with closed backs. The center diamond is likely backed with green foil. The French saying *mon cœur est à vous* is styled in gold and translates to "my heart is yours." *Ring from GeorgianJewelry.com; photos by Zachary Mial*

This Georgian pendant necklace, circa early 19th century, is crafted in darkly patinated silver over gold and set with rose-cut diamonds. *LangAntiques.com; photo by Cole Bybee*

Both women and men wore jewelry during the Georgian era. Men even wore jeweled buttons on their coats and jeweled buckles on their shoes. Women had full or partial ensembles of jewelry with matching necklaces, earrings, brooches, rings and bracelets. Full ensembles were called parures and partial ensembles were called demiparures. One unusual type of jewelry was the aigrette, a hair ornament or hat brooch that was designed to support a jeweled feather or an actual feather, from the French word for egret.

These Georgian drop earrings, circa late 18th century or early 19th century, are hand fabricated in silver over 9-karat gold and set with rose-cut and table-cut diamonds. The earrings can be worn with the long drops or with the tops alone. *LangAntiques.com; photo by Cole Bybee*

This Georgian snowflake diamond ring, circa 1750–1780, was most likely a brooch in its previous form. The one-of-a-kind snowflake design is encrusted with 5.50 carats of rose-cut and old mine-cut diamonds ranging in color from J to M. All of the gems are set with closed backs into silver tops. The shank is 18-karat white gold. *Ring from GeorgianJewelry.com; photo by Zachary Mial*

This Georgian 18-karat yellow and red gold diamond pendant from 1820 is believed to be of Belgian origin. It is set with 11 simplified rose-cut diamonds

This type of pendant is typically Belgian and dates from the late 18th to the mid-19th centuries. It is called "croix à la Jeannette" (which translates to "cross in the style of Jeanette"), a pendant in the form of a heart with a Latin cross dangling from it.

The intricate lace-like metalwork in this piece is called filigree. It is created with fine twisted or round wires and sometimes tiny beads that are soldered together in a heavier wire frame or onto a flat base. Filigree was a popular design element of European jewelry of the 17th and 18th centuries. *Pendant and photo from Adin Fine Antique Jewellery (AntiqueJewel.com)*

A Georgian table- and rose-cut diamond ring. All the diamonds are set closed backed so that from the reverse you see only gold. This ring is typical of Spanish and Portuguese designs, and it hails from the mid- to late 18th century. *Ring from GeorgianJewelry.com; photo by Zachary Mial*

A Victorian cobalt enamel locket ring from the mid-19th century. This hand-fabricated ring features a central pair of bright-white antique cushion-cut diamonds amid an interlaced design of bright, glossy cobalt-blue enamel. An array of bright-white diamonds glitter in between the woven enameled ribbons. *LangAntiques.com; photo by Cole Bybee*

## VICTORIAN (1837–1901)

Queen Victoria was a diamond lover, and she was the trendsetter for the wealthy people of Britain's new industrial society. In the early Victorian period, most diamonds were rose cuts or old mine cuts, but by the end of the 19th century these were outnumbered by the old European cut, which some people called the Victorian cut. Enameling was sometimes used as a dramatic background for the diamonds.

In the mid- and late Victorian periods, diamonds were plentiful in jewelry, especially after the discovery of diamonds in South Africa in 1867 and the introduction of electric lighting in the 1880s, which helped diamonds sparkle indoors and at night. Closed and foiled settings were gradually replaced with open-back mountings, and there was a greater variety of setting styles: bezel, prong, gypsy and wirework settings. In 1886 Tiffany & Co. introduced a high-prong setting for diamond solitaires that became the standard for engagement rings; it was appropriately called the Tiffany setting.

Gold was also readily available during the Victorian period, thanks to discoveries in California (1848), Australia (1851), Black Hills in South Dakota (1874), South Africa (1886), Yukon (1895) and Alaska (1898). Most early Victorian diamond jewelry was handcrafted using 18- to 22-karat gold, some of it tricolor (yellow, white and rose), but in 1854 the British government made 9-, 12- and 15-karat gold legal in order to meet foreign competition. Jewelers in Britain were not required to mark their jewelry during most of the 19th century, so it is not uncommon for jewelry from this period to be unmarked. By the end of the Victorian era, a high percentage of gold jewelry was machine made and mass-produced. Platinum jewelry had also been introduced to the market, but it was generally made by hand.

Victorian jewelry displayed a wide variety of motifs: branches, shells, knots, buckles, plants, flowers, vines, insects, birds, crosses, hearts, snakes, clasped hands and women with flowing hair. Amethyst, coral, garnet, turquoise and seed pearls were often used in Victorian jewelry. After the death of Queen Victoria's husband, Prince Albert (1819–1861), mourning jewelry with black onyx or jet became popular.

Many of the world's most famous jewelry firms were founded during the Victorian period: Tiffany & Co. in 1837, Cartier in 1847 and Boucheron in 1858.

Two Victorian Swiss lapel watches. Left and center are the front and back views of a diamond and sapphire, silver-topped gold lapel watch. The diamonds are rose cuts with a total weight of about 3.70 carats. The dial has black Arabic numerals with calibrated outer gold dot minute markers, gold "Louis XVI" hour and minute hands and an acrylic crystal. On the right is a lady's rose-cut diamond and 14-karat gold lapel watch and a matching crown pin also set with rose-cut diamonds. *Photos © Heritage Auctions (HA.com)*

A Victorian pin adorned with a painted and enameled miniature, diamonds and silver over 18-karat red gold, circa 1850. The graceful asymmetrical garland, with 73 rose-cut diamonds, frames a delicate hand-painted enameled miniature depicting a woman sitting barefoot in the grass petting a dog.

The brooch's style and theme are typical of the early Victorian era, which covers the coronation of Victoria as Queen of Great Britain and Ireland and her marriage to Prince Albert. *Pin and photo from Adin Fine Antique Jewellery (AntiqueJewel.com)*

A Victorian diamond, sapphire, ruby and enamel gold-hinged bangle bracelet. The old European-cut diamonds are set in 18-karat white gold surrounded by blue enamel on 18-karat gold. *Photo © Heritage Auctions (HA.com)*

A Victorian marquise-shaped old mine-cut diamond ring, circa 1870. The marquise shape elongates the finger. Unlike old European cuts, which are round, old mine-cut diamonds have an irregular perimeter; they are not quite square nor rectangular and not quite round. *Ring from GeorgianJewelry .com; photo by Zachary Mial*

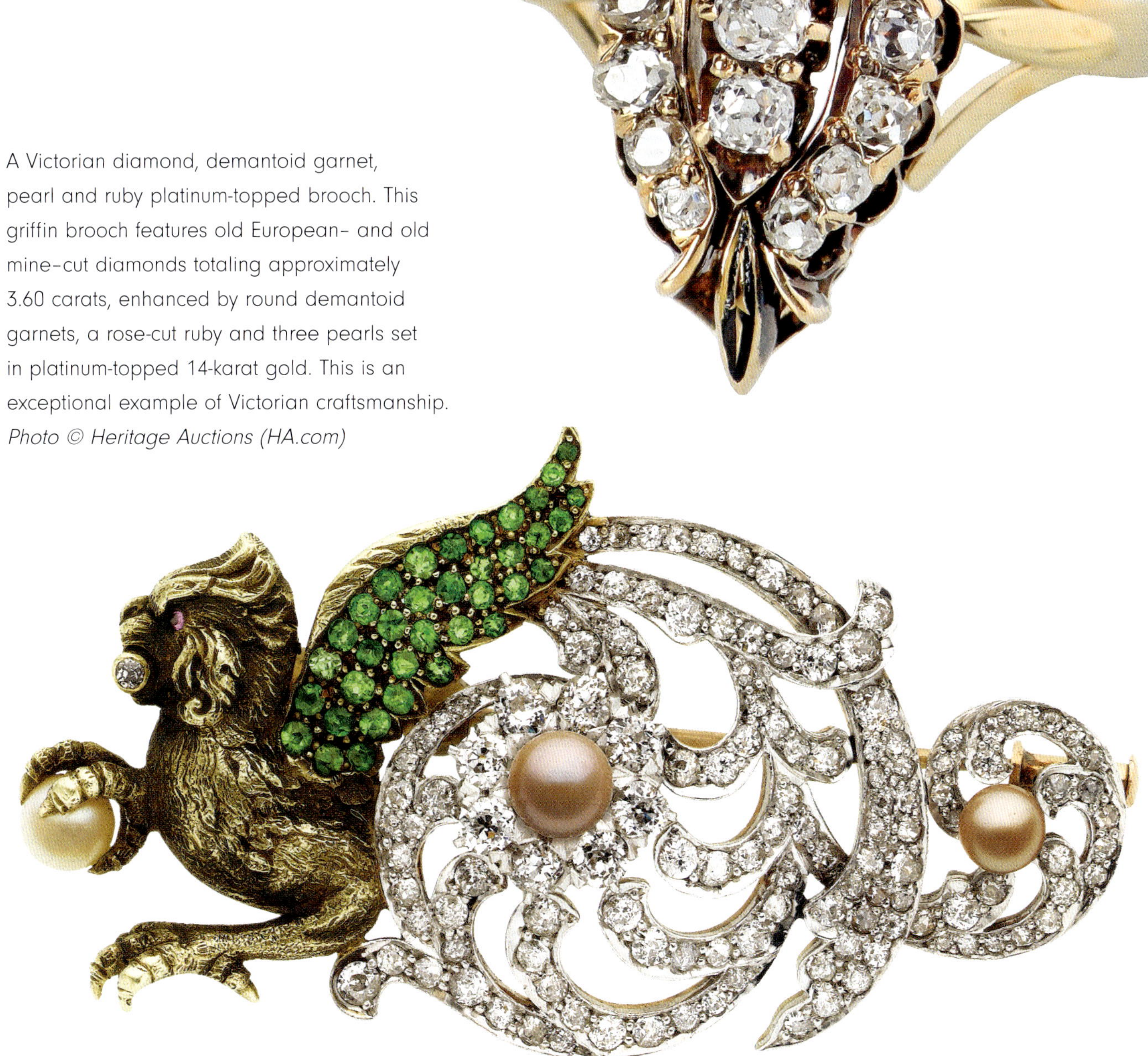

A Victorian diamond, demantoid garnet, pearl and ruby platinum-topped brooch. This griffin brooch features old European- and old mine-cut diamonds totaling approximately 3.60 carats, enhanced by round demantoid garnets, a rose-cut ruby and three pearls set in platinum-topped 14-karat gold. This is an exceptional example of Victorian craftsmanship. *Photo © Heritage Auctions (HA.com)*

A Victorian diamond, emerald, ruby and cultured pearl gold brooch. This pierced brooch features one old mine–cut diamond and one pear-shaped diamond enhanced by European-, mine- and rose-cut diamonds, pear-shaped and oval emeralds, cushion-shaped and rectangular rubies and pearls set in 18-karat gold. *Photo © Heritage Auctions (HA.com)*

A late Victorian suffragette diamond ring, circa 1900, set with an amethyst and emerald in 18-karat gold with silver. Together the colors represent a historical secret message: "Give (green) Women (white) the Vote (violet)." The three hues also represented qualities that the suffragettes aspired to: hope, purity and nobility/dignity. Women were encouraged to wear the colors to show their support for the suffragette movement and to promote public awareness of it, but it has also been suggested that these women used this secret color code out of fear of revealing their sympathies to their husbands and sons. The Fancy yellow old mine-cut diamond underlines the international common cause of women fighting for their right to vote. *Ring and photo from Adin Fine Antique Jewellery (AntiqueJewel.com)*

A Victorian fancy color diamond and silver-topped gold ring. It features a Fancy yellow rose-cut diamond encircled by rose-cut diamonds set in silver-topped 14-karat yellow gold. *Photo © Heritage Auctions (HA.com)*

A Victorian black onyx locket pendant with a diamond-loaded bouquet, made circa 1870 and probably of Belgian origin. It is set with 34 rose-cut diamonds and small diamond chips. The pendant is silver on top backed with 18-karat rose gold and has a bouquet theme inspired by mother nature. *Pendant and photo from Adin Fine Antique Jewellery (AntiqueJewel.com)*

A late Victorian French gold pendant on a chain, circa 1890, set with two old mine-cut diamonds, one small rose-cut diamond (above the bottom pearl), one sapphire and pearls. It has the control mark for 18-karat gold representing an eagle's head that was used in France from about 1838. *Pendant and photo from Adin Fine Antique Jewellery (AntiqueJewel.com)*

A late Victorian double-serpent snake ring, circa 1890, set with an old European-cut diamond, four small rose-cut diamonds and a ruby in 18-karat yellow gold. The serpent is one of the oldest and most widespread symbols and is represented in many different cultures and religions. Snakes serve as both positive symbols, such as fertility or healing, and negative symbols, such as evil or indifference. *Ring and photo from Adin Fine Antique Jewellery (AntiqueJewel.com)*

## ART NOUVEAU (1890–1914)

Art Nouveau ("New Art") jewelry was created in France, and this movement sought to modernize jewelry design and change the classical and historical styles that had previously been popular. It was also a reaction to events in French society at that time, including women's fights to secure more rights for themselves through education and jobs. Women are common motifs in Art Nouveau jewelry.

The New Art period is known for its flowing curved lines, botanical motifs and bright colors. It was the beginning of modern jewelry design. Many French jewelers adopted this style, but the one best known for his designs and artistry was René Lalique. He combined expensive gems with inexpensive materials like ivory and horn and set them in 18-karat gold. One of the techniques Lalique is noted for and that became associated with Art Nouveau is "plique-à-jour" (French for "letting in daylight") enameling — translucent enamel with no metal backing that resembles a stained-glass window.

In contrast to earlier periods, precious stones like diamonds, rubies, emeralds and sapphires were used mainly as accents for larger cabochon-cut semiprecious stones, such as lapis lazuli, moonstone, malachite, carnelian, marcasite and conch pearls. Synthetic rubies and emerald triplets (two pieces of colorless beryl joined with a layer of green cement) made their appearance in Art Nouveau jewelry. The motifs most frequently seen are women with flowing hair, human forms with insect wings, butterflies, peacocks, bees, swans, snakes and flowers. Silver, gold, copper and plated metals were all used in this type of jewelry. The Art Nouveau movement spread to other European countries as well as the United States.

Art Nouveau jewelry pieces, circa 1900. The pendant/brooch on top was hand fabricated in softly textured 18-karat yellow gold and platinum. The old mine-cut diamond at the center is highlighted with dainty platinum leaves set with rose-cut diamonds among clusters of pink-enameled flowers. A freshwater pearl drop anchors the scrolling gold work.

On the bottom, this French necklace with an asymmetrical floral design was hand fabricated in softly textured 18-karat yellow gold and set with a 1-carat European-cut diamond and a matching 0.85-carat diamond swinging below. *LangAntiques.com; photos by Cole Bybee*

An Art Nouveau brooch made with diamonds, demantoid garnets, plique-à-jour enamel and platinum-topped 14-karat gold. The European-cut diamond weighs about 0.95 carats. The piece is marked AH for August Wilhelm Holström (1829–1903), a senior member and head jeweler of Fabergé's workshop. *Photo © Heritage Auctions (HA.com)*

An Art Nouveau pendant/brooch with plique-à-jour enamel, circa 1890, showing a nymph gazing at two shooting stars made of old mine-cut diamonds amid a pink and yellow enameled dusk. A pearl embellishes a pair of green and blue plique-à-jour enameled wings as they embrace this blossom-adorned maiden. Although it does not carry any legible control marks, the pendant is believed to be of French origin. Thanks to a discreetly placed ring system in the back, it can be worn either as a brooch or as a pendant. *Pendant/brooch and photo from Adin Fine Antique Jewellery (AntiqueJewel.com)*

A pair of Art Nouveau Swiss pendant watches. Left and center are the front and back views of a dragonfly-themed 18-karat gold and enamel watch set with diamonds, emeralds, rubies and pearls. The dial has a white enamel face, gold Arabic numerals, gold filigree hands and a glass crystal. On the right is an 18-karat gold, enamel and diamond watch, circa 1905, with a 10-karat gold pendant pin. The movement and case are signed Haas Neveux & Co. *Photos © Heritage Auctions (HA.com)*

An Art Nouveau diamond, emerald and ruby 18-karat gold bracelet by Lebolt & Co., a Chicago jewelry, holloware and flatware manufacturer. The diamonds are European cuts. Engraved on the clasp: L.T.A. Aug. 18 '08. *Photo © Heritage Auctions (HA.com)*

This Arts and Crafts piece is a two-tone 14-karat gold ring set with a 7-carat moonstone and two round diamonds. The maker's mark indicates it was made by Allsopp Brothers of Newark, New Jersey. *LangAntiques.com; photo by Cole Bybee*

## ARTS AND CRAFTS (1890–1914)

The Arts and Crafts jewelry movement originated in Britain and coincided with three other schools of design — late Victorian, traditional Edwardian and Art Nouveau. It was a reaction to the mass-produced jewelry of the industrial revolution and the lavish ornamentation of the Victorian era.

Arts and Crafts jewelry was entirely handmade, and the materials were usually inexpensive. Silver, brass and copper were the preferred metals instead of gold and platinum. Cabochon turquoise, agate, moonstones, amber and opals often replaced diamonds, rubies and emeralds and were usually bezel set (a bezel setting uses a strip or strips of metal to secure a gem in place, as opposed to prongs). The designs were either abstract or featured motifs of nature, such as flowers, leaves and birds.

## EDWARDIAN (1890–1915)

The heavy use of diamonds, platinum and pearls in delicate, lacelike mountings are chief characteristics of the Edwardian era of jewelry. Even though the reign of Edward VII was only from 1901 to 1910, the lavish court style of the era influenced fashion in the decades before and after his rise to power.

Edwardian jewelry was also inspired by the French courts of Louis XV and Louis XVI. In fact, it was the French jeweler Louis Cartier (1875–1942) who was at the forefront of developing this style and who was the official jeweler for the English court. As a result, it has also been identified as the Belle Epoque, a French term meaning "beautiful era." Another expression that is sometimes used is *style guirlande*, or the "garland style," because garlands (wreaths) of flowers and leaves were typical motifs. Other motifs included horseshoes, doves, ducks, fish, hearts, sun, stars, moon, and bow knots and arrows.

New cuts such as the marquise, the emerald and the baguette emerged during this period, thanks to improvements in diamond-cutting technology. Calibrated stones of standardized sizes and shapes became available for use in mass-produced jewelry. Cushion

An Edwardian demantoid and diamond brooch, circa 1900, set with 82 Russian demantoid garnets and 143 old European-cut diamonds in a mounting of platinum over 18-karat yellow gold. *Bird brooch from GeorgianJewelry.com; photo by Zachary Mial*

cuts, old European cuts and rose cuts remained popular, and briolette diamonds were often suspended in earrings.

After the development of a torch hot enough to work it, around 1890, platinum became the most common metal for fine pieces. For a while, platinum was laminated to gold, much like silver had been. Gradually it became evident that platinum was strong enough to be used by itself for intricate mountings and secure diamond settings. Platinum milgrain (raised beaded edges) settings were used to make diamonds look larger, and knife-edge settings were created to make the mounting appear invisible. Much of the metalwork was open, allowing fabric to show through. During World War I white gold came into common use because platinum was temporarily banned for use in jewelry because it was needed for the war effort.

Even though Edwardian jewelry was primarily white, pastel colors were also in fashion, and later pieces were set with darker-colored gemstones, such as amethyst, alexandrite, chrysoprase and demantoid garnets. Some of the noted French houses who created Edwardian jewelry were Cartier, Boucheron, Chaumet, Georges Fouquet and LaCloche Frères. In the United States Edwardian jewelry was made by Tiffany & Co., Black Starr & Frost, Marcus & Company and others. Most of the pieces of the Russian imperial court jeweler Peter Carl Fabergé (1846–1920) can be classified as Edwardian, but other creations by Fabergé have Art Nouveau lines and motifs.

An Edwardian diamond ring, circa 1910, set with old European-cut diamonds in platinum over 18-karat gold. The delicate lacy platinum metalwork is typical of the Edwardian era. *Ring from GeorgianJewelry.com; photo by Zachary Mial*

An Edwardian emerald, diamond, platinum and gold brooch. This stunning piece features a Colombian emerald weighing approximately 7 carats, enhanced by European-cut diamonds and square-cut emeralds set in platinum and 18-karat gold. *Photo © Heritage Auctions (HA.com)*

An Edwardian aquamarine, diamond, platinum and gold pendant and brooch. It features an oval aquamarine weighing approximately 13 carats, enhanced by rose-cut diamonds weighing a total of approximately 1.60 carats and accented with European-cut diamonds. The piece is set in platinum and complemented by a 14-karat gold pinstem and catch. It features a delicate-looking bow motif as well as a fine milgrain finish. *Photo © Heritage Auctions (HA.com)*

This Edwardian diamond and platinum pendant and brooch features a European-cut diamond weighing about 0.90 carats, which is enhanced by European- and single-cut diamonds. *Photo © Heritage Auctions (HA.com)*

An Edwardian pearl, diamond and platinum pendant necklace, circa 1910. This intricate peacock plume pendant is set with 60 natural pearls, 87 old mine-cut diamonds and 37 rose-cut diamonds. The fineness of line and lacey aspect of the piece are typical of the Edwardian period. Although the seller, Adin, cannot say with certainty who made this piece, there are strong signs that it could have been Soler Cabot, a well-known *haute joaillerie* (high jewelry) house from Barcelona, Spain. Soler Cabot was established in 1842 and is still run by the family. The Cabot dynasty began in Sant Andreu de Llavaneres, Spain. Several generations have been registered with the Royal Academy of Silversmiths in the nearby city of Mataró, dating as far back as 1662, but some documents place the origins of the dynasty around 1543.

Peacock plumage is the theme of the piece. In the Middle Ages, Christian artists sometimes used the peacock to symbolize the idea of immortality. Peacocks are also associated with the ancient Greek goddess Hera and the Hindu gods Murugan and Krishna. *Necklace and photo from Adin Fine Antique Jewellery (AntiqueJewel.com)*

This Art Deco natural Burmese jade and diamond ring was hand fabricated in platinum. The untreated jade cabochon is 0.33 by 0.24 by 0.26 inches (8.40 by 6.05 by 6.50 millimeters). *LangAntiques.com; photo by Cole Bybee*

## ART DECO (1915–1940)

Geometric patterns, straight lines, symmetry and bold color contrasts characterize Art Deco jewelry designs. Platinum, diamonds and white gold continued to be used extensively, but there was a greater use of colored gemstones, some of which were synthetic. Lapis, jade, coral and black onyx were especially popular.

Even though old European, single and rose cuts were still present, the modern brilliant cut in round and fancy shapes was more commonplace. New shapes for side stones emerged in the form of bullets, half-moons and shields.

One of the most popular articles of jewelry was the diamond straight-line bracelet, which was revived in the 1980s and called the tennis bracelet. The cocktail wristwatch, pendant watch and dress clip were also in vogue during the Art Deco period.

The motifs most often found were geometric, abstract, floral and inspired by aesthetics from ancient Egypt, China, Japan and India, among other cultures. Art Deco Egyptian and Japanese motifs followed the celebrated discovery of King Tutankhamun's tomb in 1923 and new trade agreements between Japan and the U.S.

Louis Cartier is the most famous Art Deco designer. The works of Van Cleef and Arpels also had a strong influence on the period. Other leading designers and houses were Mauboussin, Jean Fouquet, Boucheron, Chaumet, LaLoche and the American firms of Tiffany & Co., Black, Starr & Frost, J.E. Caldwell & Co., C.D. Peacock, Harry Winston, and Shreve, Crump & Low.

An Art Deco diamond, sapphire and platinum ring by J.E. Caldwell, a Philadelphia jewelry manufacturer known for its high-quality Art Nouveau and Art Deco jewelry. The center old European-cut diamond weighs about 1.30 carats. *Photo © Heritage Auctions (HA.com)*

An Art Deco diamond, ruby and black onyx platinum spiderweb ring. The center old mine-cut diamond is 1.64 carats and has a GIA color grade of I and a clarity grade of $VS_1$. *LangAntiques.com; photo by Cole Bybee*

An Art Deco diamond, ruby and emerald platinum bracelet. This tutti-frutti-inspired bracelet features European- and transitional-cut diamonds weighing a total of about 12 carats, carved emerald leaves, ruby cabochons and single- and rose-cut diamonds all set in platinum. The tutti-frutti jewelry style features rubies, sapphires and emeralds, often with diamond accents, and was first created by Cartier in 1901, when he made a necklace for Queen Alexandra. *Photo © Heritage Auctions (HA.com)*

An Art Deco diamond, ruby, black coral, enamel and platinum brooch by Oscar Heyman & Bros. The brooch features a carved ruby, full- and single-cut diamonds weighing a total of approximately 0.70 carats, trapezoid and baguette-cut diamonds weighing a total of approximately 0.25 carats and black coral all set in platinum. This piece was authenticated by Oscar Heyman & Bros. The Heyman family emigrated from Russia to New York in 1906, and they launched Oscar Heyman & Bros in 1912. They specialized in high-end platinum jewelry set with diamonds, rubies, sapphires and emeralds. *Photo © Heritage Auctions (HA.com)*

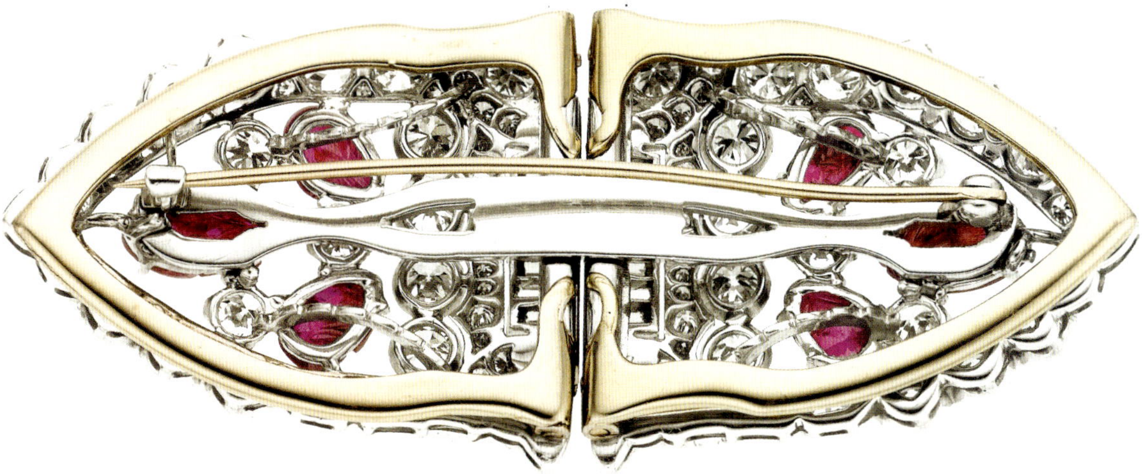

An Art Deco diamond, ruby and platinum clip and brooch. This piece features European-, full- and single-cut diamonds weighing a total of approximately 7 carats, which are enhanced by carved rubies, baguette-cut diamonds and baguette-cut rubies all set in platinum. It is completed by a 14-karat white gold clip mechanism. *Photos © Heritage Auctions (HA.com)*

An Art Deco diamond, ruby and platinum ring by Oscar Heyman & Bros. This ring features a 1.05-carat rectangular step-cut diamond and full- and single-cut diamonds, which are accented by baguette-, square- and calibré-cut rubies all set in platinum. *Photo © Heritage Auctions (HA.com)*

An Art Deco black onyx and diamond bow brooch, circa 1930. Black against white became one of the hallmarks of the 1920s and much of the '30s, during the Art Deco era. The diamonds are a blend of cuts – triangular (trillion or trilliant), old European, old single and square shapes – set in platinum with milgrain beaded edges. *Bow brooch from GeorgianJewelry.com; photo by Zachary Mial*

An Art Deco diamond, sapphire and platinum ring. It features a marquise-shaped modified brilliant-cut diamond weighing 3.55 carats, which is enhanced by calibré-cut sapphires and European-cut diamonds weighing a total of approximately 0.75 carats. The piece is set in platinum. *Photo © Heritage Auctions (HA.com)*

This Art Deco diamond, glass and platinum ring features a European-cut diamond weighing approximately 2.70 carats, which is enhanced by European- and mine-cut diamonds that weigh a total of approximately 0.30 carats. The stones are set in platinum and accented by an Art Deco–style contrasting square frame of green glass. *Photo © Heritage Auctions (HA.com)*

A Retro diamond, sapphire, platinum and 14-karat gold clip brooch. The diamonds are in a pavé setting. For this type of setting, the diamonds are fitted into tapered holes and set almost level with the surface of the brooch. Some of the surrounding metal is then raised to form beads that hold the diamonds in place. This type of setting was often used during the Retro period to create large bright and bold surfaces. *Photo © Heritage Auctions (HA.com)*

**RETRO (1939–1950)**

The all-white look of diamonds and platinum began to lose its appeal during the Great Depression (1929–1939). Then when the U.S. government declared platinum a strategic metal during World War II, it was no longer used as a jewelry metal in America. It was replaced in fine jewelry by yellow gold and rose gold and later by white gold. Colored gems such as citrine, aquamarine and tourmaline were often used in addition to rubies, sapphires and emeralds. Masses of baguette-cut stones were used in jewelry that was channel set (a setting style in which gems are suspended in a channel of metal walls with no metal between them).

Hollywood stars were increasingly influencing fashion more than royalty, and France was no longer the jewelry design center of the world. Retro designs were bold, futuristic, colorful and three-dimensional. Bracelets had heavy links; pendants were large and designed so that they could convert to brooches. The main themes of the Retro period were feminine, patriotic and industrial. Common motifs included flowers, birds, bows, scrolls and belt buckles.

In 1948 DeBeers launched its famous marketing slogan "a diamond is forever." Platinum returned after the war and was used to create lighter-weight jewelry and wire settings that held clusters of diamonds, many of which had pear, oval and marquise shapes. By the end of the Retro period, diamonds were once again a girl's best friend.

A Retro diamond, aquamarine, ruby, 18-karat gold and platinum brooch by Tiffany & Co. The brooch features channel-set baguette- and bullet-cut diamonds weighing a total of about 2 carats, which are set in platinum and enhanced by round aquamarines and round rubies. *Photo © Heritage auctions (HA.com)*

This Retro diamond and lab-grown ruby rose-gold ring features a European-cut diamond weighing approximately 0.65 carats, which is accented with single-cut diamonds and synthetic ruby cabochons, all bezel set in platinum-topped 14-karat rose gold. Synthetic ruby was often used in jewelry during World War II because natural ruby was not readily available. *Photo © Heritage Auctions (HA.com)*

A Retro platinum and 18-karat gold double-clip brooch channel set with citrines and diamonds. This type of brooch consists of two matching dress clips held together by a large framework or pin that can be detached to form two smaller brooches. Double-clip brooches were first created by Cartier in the 1920s. *Photo © Heritage Auctions (HA.com)*

A Retro diamond, ruby, enamel and gold pendant and brooch. This rose-shaped brooch features a round brilliant-cut diamond weighing about 0.80 carats, which is surrounded by single-cut diamonds weighing a total of approximately 2 carats. The stem is accented by square channel-set rubies weighing a total of approximately 0.40 carats, with applied enamel for the leaves. The whole piece is set in 14-karat white and pink gold. *Photo © Heritage Auctions (HA.com)*

A Retro sapphire and diamond 18-karat gold French bracelet. The hinged bangle features square and baguette-cut sapphires, which are enhanced by European-, single- and rose-cut diamonds set in gold. *Photo © Heritage Auctions (HA.com)*

A Retro pearl and diamond platinum ring. Designed as an oyster shell, it features a medium-sized cultured pearl enhanced by round brilliant-cut diamonds set in platinum. *Photo © Heritage Auctions (HA.com)*

A Retro ruby, diamond and cultured pearl 18-karat gold brooch. This brooch, depicting a Spanish dancer, features single-cut diamonds enhanced by square-cut rubies and a rectangular synthetic ruby. The dancer's head is a medium-sized cultured pearl, and the whole piece is set in 18-karat gold. *Photo © Heritage Auctions (HA.com)*

A Retro diamond and ruby 14-karat gold buckle ring, circa 1945. Buckles were all the rage in the 1940s for all types of jewelry, from bracelets to brooches. The frame of the buckle is set with four single-cut diamonds, the tongue with two synthetic rubies and the belt tip with seven channel-set synthetic rubies. Smooth to the touch, the mesh band is flexible. The adjustable ring has a double catch to secure the mesh "belt" and was designed to fit most anyone. *Ring from GeorgianJewelry.com; photo by Zachary Mial*

A Retro diamond, ruby, platinum and gold jewelry suite by Merrin. The suite includes a pair of earrings featuring rubies enhanced by brilliant- and single-cut diamonds set in platinum and completed with clip backs. The matching brooch features rubies weighing a total of about 25 carats, accented by full- and single-cut diamonds, baguette-cut diamonds and marquise-cut diamonds. Both earrings and the brooch are set in platinum. Merrin was a New York-based design firm. *Photos © Heritage Auctions (HA.com)*

A Mid-Century matching diamond brooch and pair of earrings from the 1960s. This custom-crafted set is made of platinum and 18-karat yellow gold. *LangAntiques.com; photo by Cole Bybee*

### MID-CENTURY (1950–1970)

Following World War II, as diamonds became more available, jewelry emphasizing diamonds became more widespread. Mid-Century jewelry was often set with baguettes and fancy-shaped diamonds. Consumers had a choice of affordable jewelry with a single small diamond or elaborate pieces made up of multiple diamonds with swirling and cascading designs. Mid-Century jewelry also brought back the popularity of matching jewelry sets that had a single motif through a necklace, brooch, bracelet and set of earrings.

After the U.S. lifted the war embargo on platinum, it was used in high-end jewelry once again. However, platinum did not regain its previous popularity because white gold was less expensive and easier to work with, and the public had readily accepted the change of metal.

A notable change in Mid-Century jewelry was a move away from high-polished metal to a more textured appearance with brushed finishes, mesh and braided metal. Necklaces tended to be shorter and were sometimes used as hair ornaments. Large rings were in style, and charm bracelets were big sellers. Earrings came in both button style and dangling, and large ear clips were available in a wide variety of shapes and designs. Common motifs included bows, animals, insects, fish, nuts, berries, starbursts, leaves, single flowers and flower bouquets. Overall, Mid-Century jewelry tended to be more feminine and elegant than the bold styles of the Retro period.

A Mid-Century platinum flower basket brooch, circa 1950, set with transitional-cut diamonds and emeralds in a variety of shapes, from marquise and pear to round diamonds. Milgrain (beaded edge) details outline the intricate metalwork. *Brooch from GeorgianJewelry.com; photo by Zachary Mial*

A one-of-a-kind Mid-Century 18-karat yellow and white gold diamond solitaire ring, circa 1960. The 0.12-carat diamond is set in a white gold floral-like setting, which increases its apparent size. On the back of the band is an eagle's head hallmark, which indicates this unique ring is of French origin. *Ring from GeorgianJewelry.com; photo by Zachary Mial*

A Mid-Century diamond half eternity band, circa 1970. The marquise diamonds alternate with opposing tapered baguettes in a platinum mounting. *Ring from GeorgianJewelry.com; photo by Zachary Mial*

A Mid-Century diamond and platinum ring with a bow motif. This diagonally oriented ring features a 0.80-carat European-cut diamond, set with round brilliant-cut diamonds and diamond baguettes. *LangAntiques.com; photo by Cole Bybee*

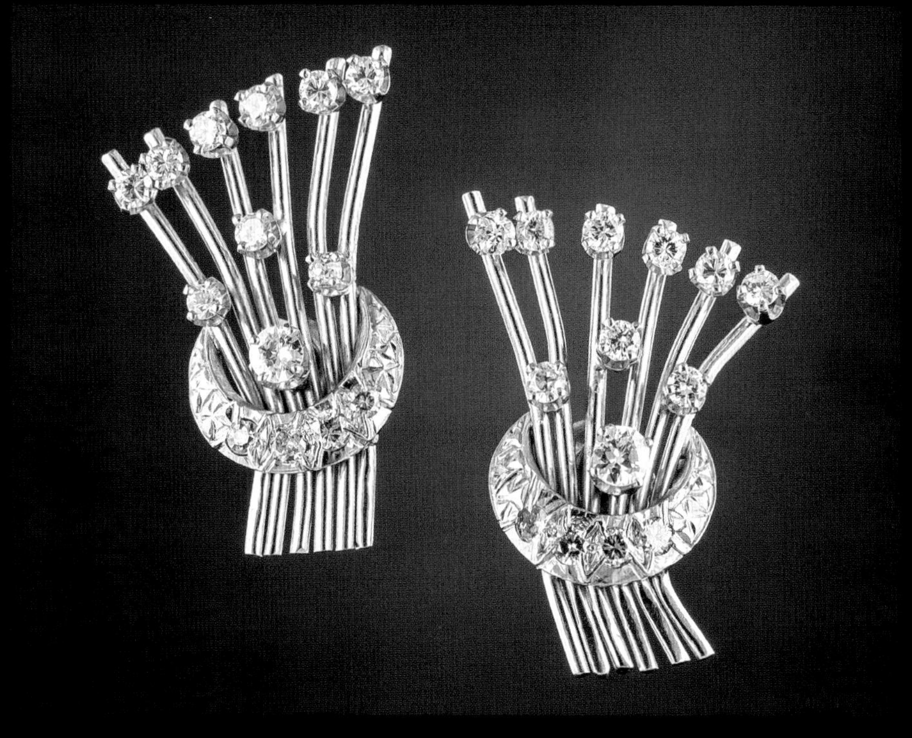

A pair of Mid-Century diamond spray earrings resembling shooting stars, circa 1950s. They are set with modern round brilliant-cut diamonds set in handcrafted 14-karat white gold. *LangAntiques.com; photo by Cole Bybee*

Mid-Century natural jadeite earrings set in platinum with swirls of tapered baguette diamonds and cascades of round and marquise-shaped diamonds. *LangAntiques.com; photo by Cole Bybee*

Mid-Century unheated pink sapphire and 14-karat white gold ring set in a swirl of baguette-cut and round brilliant diamonds. *LangAntiques.com; photo by Cole Bybee*

East-west marquise-shaped diamond ring. *Ring and photo by Peter Indorf*

Diagonally set marquise-shaped diamond ring. *Ring and photo by Glenn Lehrer*

## MODERN (1970–PRESENT)

The Modern jewelry era is a composite of many jewelry styles, including those that existed well before 1970. In the early 2000s, for example, halo rings became the leading engagement ring style. They feature a central gemstone encircled by a "halo" of smaller diamonds or other gems. This setting is not only attractive, but it also makes the center stone appear larger. This is far from a new style, however. Its origins can be traced back to the early Georgian era, when slightly smaller round diamonds or pearls were set around a larger center stone. Modern halo rings are influenced by the Art Deco jewelry of the 1920s, with dramatic, large center stones often surrounded by smaller stones. Since then, halo settings have gone in and out of fashion, and between 1970 and 2000, halo rings were seldom promoted. Baguette diamond rings were common in the 1980s, whereas solitaires and three-stone diamond rings were more frequently marketed in the 1990s.

Every year jewelry magazines and organizations discuss the newest jewelry trends. In 2020 the Natural Diamond Council listed the following jewelry trends:

- diamonds set in colorful enamel and ceramic designs;
- personalized jewelry with the initials and names of the wearer;
- long, dangling statement earrings;
- organic and earthy commitment rings set with natural rough diamonds;
- mismatched earrings;
- layered necklaces;
- horizontally or diagonally set marquise-, oval- and emerald-cut diamonds in "east-west rings."

Fashion trends are fleeting, so readers who want to learn what is trending in diamond jewelry today should visit the Natural Diamond Council website (NaturalDiamonds.com). Ultimately what is most important is that consumers select styles that flatter them and the gems they purchase, not whether the piece is the current vogue.

Diamond halo rings in yellow, white and rose golds. The halo ring has been one of the most popular engagement ring styles of the 21st century. *Diamond bridal rings from the Satin Collection of John Atencio; photo courtesy of John Atencio*

The biggest changes that have occurred in diamond jewelry since 1970 involve the diamonds themselves. Low-grade diamonds that had previously been considered unacceptable in fine jewelry are now being promoted in high-end designer jewelry. For example, diamonds in Edwardian, Art Deco, Retro and Mid-Century jewelry were always transparent, but after 1970 cloudy, translucent diamonds started appearing in mass-produced, low-budget jewelry. Today they are sometimes sold as "Fancy white" diamonds in high-priced jewelry. By the early 2000s semiopaque rough diamonds could be found in designer jewelry, and in 2020 advertisements appeared for "salt and pepper diamonds" — with many black, gray and white inclusions — which had previously been considered industrial-grade diamonds.

After the Argyle mine in Australia opened in 1983, brown and light brown diamonds became widely available. The mine was also noted for its high-quality pink to red diamonds. Discoveries at other mines

also helped increase the supply of colored diamonds. The branding of unique diamond cutting styles became more widespread in the 1990s, and diamonds colored by high pressure and high temperature (HPHT) treatment were introduced to the market in 2000. By 2020 natural, treated and lab-grown diamonds of many colors and cutting styles were being used in jewelry.

In the early 1990s platinum began to reemerge as an important jewelry metal in North America. Japanese consumers, however, were already aware of its advantages. About 90 percent of all wedding and engagement rings sold in Japan were being made of platinum. Interestingly platinum had not been banned in jewelry during World War II in Japan. Instead, the Japanese government had considered gold to be a strategic metal.

In 1992 Platinum Guild International USA (PGI) was formed for the purpose of reeducating American consumers about the durability and elegance of platinum. Technical training was offered to jewelers to help them deal with the challenges of working with platinum. PGI's efforts paid off. Platinum jewelry is now widely available, and many Americans are enjoying its benefits. Its popularity has also spread to China and India.

As for shapes, round diamonds remain in style, whereas demand for fancy shapes changes from one decade to another. For example, in the 1980s marquise cuts were in high demand and could cost more than rounds of the same size and quality. Later, princess cuts became more popular. However, today some dealers say it is difficult to find buyers for certain fancy cuts. According to the February 2021 issue of *Rapaport* magazine, oval diamonds were in high demand and difficult to find, and pears and cushions were in good demand. But no matter what shape or cutting style a customer might want, it is available and used in Modern jewelry — even old cuts, such as table cuts, rose cuts and old mine cuts. One of the first to use diamond rough and old-cut diamonds in contemporary luxury jewelry was designer Todd Reed of Boulder, Colorado.

(opposite page) Hand-forged diamond rings by jewelry designer Todd Reed. These rings are set with natural-cut and rough diamonds in a wide variety of colors, shapes, cutting styles and transparencies. Sterling silver, palladium, 18-karat gold and sterling silver with black patina are the metals used in these rings. *Photos courtesy of Todd Reed*

This pendant, entitled Baag, is a Siberian white tiger. The piece is set with white, brown and black diamonds, two aquamarines for the eyes and one Oregon fire opal for the nose. (*Baag* or *Bagh* means "tiger" in Hindi and Himalayan languages.)

This piece, entitled Okami, which is "wolf" in Japanese, is encrusted with brown diamonds, white diamonds and labradorite and completed with amber eyes and a black jade nose. The gray wolf has been wiped out in most of its historical range in Europe, Asia and North America. In places where they have been reintroduced, like Yellowstone Park, there have been unanticipated enrichments to the variety and abundance of other species of plants and animals.

This Crevoshay pendant, entitled Clouded Leopard, is made of 18-karat yellow gold and encrusted with black and brown diamonds. The nose is made of Oregon fire opal and the eyes of aquamarine and black diamond.

This giraffe pendant-brooch, entitled April, is made of 18-karat yellow and rose gold and adorned with brown and black diamonds and two emerald eyes.

Pendants and brooches by jewelry artist and painter Paula Crevoshay. These realistic portrayals of animals are part of Crevoshay's Endangered Species Collection. Her work has been exhibited in many museums and is on display at the Gemological Institute of America, the Carnegie Museum of Natural History and the National Gem Collection at the Smithsonian Institution. *Photos by Crevoshay Studio*

Long, dangling statement earrings, like the 18-karat rose gold earrings above, were a trend in 2020. The button-style earrings on the left are for people who prefer a simpler style.

Both pairs of earrings were designed and made by the Asian Star Group, and all the diamonds in them were cut and polished by the same company. *Photos © Asian Star Group*

A diamond bracelet set with round brilliants and translucent colored diamonds by Hubert Jewelry. *Photo by Diamond Graphics*

Rough and faceted diamonds in a ring and earrings by Hubert Jewelry. *Photos by Diamond Graphics*

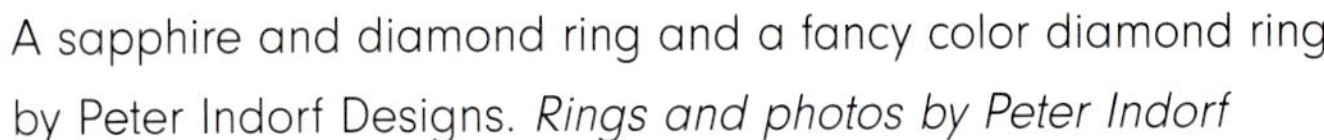

A sapphire and diamond ring and a fancy color diamond ring by Peter Indorf Designs. *Rings and photos by Peter Indorf*

These fancy color diamond rings feature pink diamonds from the Argyle diamond mine in Australia, which was established in 1983 and closed in 2020. Argyle produced more pink and brown diamonds than any other mine during that period. The ring on top is a combination of the halo and *toi et moi* (French for "you and me") styles. *Toi et moi* rings have one band that features two main gems, each representing a spouse, and this style dates back to the 1500s and the Victorian period. This ring style has reappeared in Modern jewelry. *Rings and photos courtesy of Gems by Pancis*

A diamond and 18-karat gold tiara cuff bracelet by Barbara Heinrich. *Photo by Tim Callahan*

A black diamond medallion and bead necklace by Barbara Heinrich. Designers began using black diamonds in their jewelry in the 1990s, and their popularity increased after a 5-carat black diamond engagement ring was presented to the character Carrie in the 2010 movie *Sex and the City 2*. *Photo by Tim Callahan*

A Brazilian Paraiba tourmaline and diamond necklace. The electric blue color, internally flawless clarity and tremendous size (nearly 60 carats) of this Paraiba tourmaline makes it one of the rarest in the world. The Paraiba gem is encircled by Fancy pink diamonds and a wreath of marquise- and pear-shaped diamonds. The pendant is suspended from a strand of more than 50 carats of diamonds. Each diamond was hand cut and calibrated to match the corresponding diamonds on the left and right. *Necklace and photo courtesy of Dehres Ltd.*

These five extraordinary diamond rings by Dehres Ltd. are truly rare and modern. Clockwise from center:

- 10.28-carat pear-shaped D-color Internally Flawless diamond
- 5.02-carat radiant-cut Fancy Vivid purple-pink $VVS_1$ diamond named Bubblegum Pink
- 16.32-carat radiant-cut Fancy Vivid yellow $VS_1$ diamond
- 7.68-carat radiant-cut Fancy Vivid pinkish-orange $SI_2$ diamond named the Lotus Bloom
- 2.84-carat cushion-cut Fancy Intense bluish-green $SI_2$ diamond

*Rings and photo courtesy of Dehres Ltd.*

# 6

# How Are Diamonds Priced?

On April 28, 1987, a 0.95-carat Brazilian diamond was sold for $880,000 at a Christie's auction in New York City. It contained two large flaws: a deep cavity in the table and another cavity at the edge. Despite its flaws and small size, it set a world record for the highest per-carat price for any gem sold at auction — $926,000 — and it held that record for 20 years. The diamond was named the Hancock Red after Warren Hancock, a collector who had reportedly bought it in 1956 for $13,500 from his local jeweler.

Blazing Red, a 0.92-carat Fancy red diamond (GIA). *Diamond and photo courtesy of Dehres Ltd.*

Why did the Hancock Red command such a high price? Its natural color is an extremely rare deep purplish red. Pure red colors like that of the Blazing Red diamond in the photo on the left can command even higher prices. Color is the most important factor in determining the price of a natural diamond. The rarer the color, the higher the price (all other factors being equal). The rarest and highest-priced diamond colors are violet and red. The lowest are brown and gray.

Colorless diamonds are a fraction of the cost of red diamonds. For example, a well-cut 0.95-carat flawless D-color round diamond could be purchased for less than $20,000, but factors other than color play key roles in the pricing of a diamond. The remaining seven basic price factors for diamonds, beyond color, are as follows:

- **Carat weight**
- **Cut quality:** proportions, finish and optical attributes (a.k.a. light performance)
- **Cutting style and stone shape**

- **Clarity:** the degree to which a stone is free from inclusions and blemishes
- **Transparency:** the degree to which a stone is clear, hazy or cloudy
- **Treatment status:** whether a stone is untreated or treated and what type of treatment it has received
- **Creator:** Whether a stone is natural or lab grown

In this chapter, you will learn how these factors affect diamond prices and how they are described on lab reports and diamond appraisals.

## Color

The most widely used color grading systems for diamonds are the two developed by the Gemological Institute of America (GIA): one is for colorless to light yellow, brown or gray diamonds, and the other is for natural diamonds with a noticeable depth of body color, which are called fancy color diamonds, fancy colored diamonds, fancies or fancy diamonds.

GIA's first color grading system was developed in the 1950s. It identifies colors with alphabetical letters ranging from D to Z+. D color is the highest and most expensive color grade in this grading system. The less color there is the higher the price, except when the color reaches a point just stronger than light yellow (a grade designated as Fancy Light yellow). As the color intensity of a Fancy Light yellow diamond increases, so does its price.

The following diagram helps explain the meaning of the GIA color grades. Brown and gray diamonds are graded on the same scale.

| D E F | G H I J | K L M | N to R | S to Z | Z+ |
|---|---|---|---|---|---|
| colorless | near colorless | faint yellow | very light yellow | light yellow | Fancy Light yellow |

GIA Gem Lab masterstones ranging from E to O colors. (Note: Do not use this photo to grade the color of diamonds. The printing and developing processes and paper color usually alter the true color of gems in photographs.) *Photo by Tino Hammid © GIA; reprinted with permission*

The nuances of color are so fine between the grades that the average consumer cannot tell the difference between a D and F color grade. However, the price difference between D and F colorless diamonds could be as much as 10 to 35 percent depending on their size and clarity. Professionals must rely on comparison stones to determine color grades, and even then it can be hard to distinguish between a D or E colorless diamond. Diamonds ranging from colorless to Z color are graded facedown on a nonreflective white background under standardized daylight-equivalent fluorescent lighting. The image below shows how the color appears through the sides and bottoms of the stones.

A 25.23-carat Fancy Vivid yellow diamond ring. It sold for $975,000 at the Dallas Fine Jewelry Heritage Auction in October 2020. *Photo © Heritage Auctions (HA.com)*

Diamonds that have a depth of color beyond Z color are graded using a second system that describes the colors with words instead of letter grades. For example, a yellow diamond beyond Z is graded as Fancy Light yellow, followed by Fancy yellow. The next color grade is Fancy Intense yellow. The highest color grade for a yellow diamond is Fancy Vivid yellow. If the yellow is masked by brown, the stone may be graded Fancy Dark yellow or Fancy Deep yellow depending on how dark the stone is. Pure yellow diamonds cost more than brownish-yellow ones. Unlike colorless to light yellow diamonds, fancy color diamonds are viewed faceup when grading them.

The diagram on the next page illustrates how color grades established by GIA relate to the amount and lightness or darkness of color (tone) and the purity of the color (saturation and the degree of brown or gray masking the color). The GIA Gem Laboratory also has masterstones that it refers to when grading fancy color diamonds. However, considering the cost of colored diamonds and the number of possible color combinations, appraisers would not be able to afford

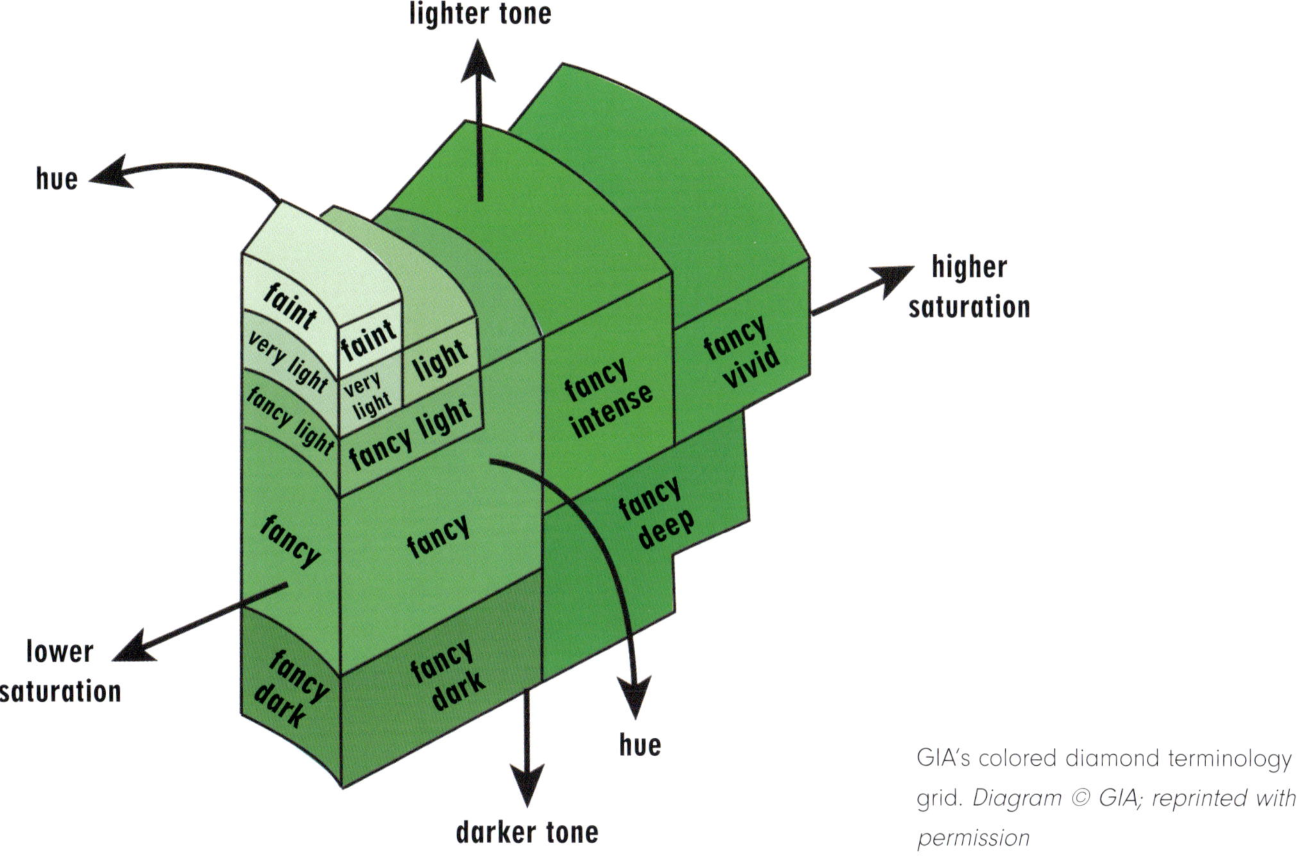


GIA's colored diamond terminology grid. *Diagram © GIA; reprinted with permission*

fancy color masterstones. Instead, they base appraisal valuations on the grades assigned by major gem laboratories with sophisticated equipment capable of determining if the color of a diamond is natural or the result of a treatment process. Only natural fancy color diamonds can receive a GIA color grade.

Pure diamond colors are rare. Secondary colors are often present and may be modified with brown or gray. The color grade "Fancy grayish-yellowish-green" is an example. A diamond graded "Fancy green" would command a higher price, with all other factors being equal. Keep in mind that the color grade represents a range of color, so two diamonds with the same grade may appear to have slightly different colors. For example, one Fancy yellowish-green diamond might appear a little greener than another diamond with the same GIA color grade.

very rare fancy color GIA-graded diamonds. (Keep in mind that the printing and developing processes usually alter the true color of gems in photographs.) *Diamonds and photos courtesy of Dehres Ltd.*

Bubblegum Pink, a 5.02-carat Fancy Vivid purplish-pink diamond.

Green Harmony, a 1.19-carat Fancy Vivid yellowish-green diamond.

Nightfall, a 1.25-carat Fancy Deep blue diamond.

The Eternal Sun, a 10.05-carat Fancy Vivid yellow diamond.

Ocean Teardrop, a 0.90-carat Fancy Vivid blue diamond.

Viva Verde, a 1.01-carat Fancy Vivid green diamond.

The GIA color grades of the center diamonds in rings by Hubert Jewelry. *Photos by Diamond Graphics*

A Fancy orangy-pink diamond.

A Fancy pink diamond flanked by Fancy blue diamonds.

A Fancy blue diamond flanked by Fancy Vivid purplish-pink and Fancy Intense purplish-pink diamonds.

A Fancy Light blue diamond.

A Fancy Intense yellowish-green diamond.

A Fancy greenish-yellow diamond.

A Fancy gray diamond.

A Fancy Light yellow diamond.

A Fancy Light green diamond.

A Fancy Deep brownish-yellow diamond.

A Fancy Intense yellow diamond.

A Fancy orangy-brown diamond.

## Carat Weight

A carat is a unit of weight equaling one-fifth of a gram (approximately 0.007 ounces). The term carat originated in the 1500s when gemstones were weighed against a carob bean. Each bean weighs about 1 carat. In 1907 carat weight was standardized in Europe and adapted to the metric system. The U.S. adopted the standard in 1913, and the U.K. adopted it in 1914. Even though retail stores generally indicate the total price of each of their diamonds, gem dealers typically quote their price per carat (per-carat price) because this makes it easier to compare the prices of stones with different weights. In most cases, the higher the carat weight category, the greater the per-carat price of the diamond because larger diamonds are more desirable; however, it's important to bear in mind that carat weight is only one factor in pricing diamonds, and other factors such as color, cut and clarity will have a great impact on the price.

The weight of small diamonds is frequently expressed in points, with one point equaling 0.01 carats. For example, five points is a short way of saying five one-hundredths of a carat. Diamonds weighing 0.05 carats are referred to as five pointers. If a diamond is advertised as weighing "0.25 points," this can be misread as being 0.25 carats (or one-quarter of a carat) in weight, when in fact 0.25 points is equal to one four-hundredth of a carat.

In diamond districts, you may also hear the term grainer. This word describes the weights of diamonds in multiples of 0.25 carats, which is one grain. Therefore, a four grainer is a 1-carat diamond.

Consumers could also misinterpret the labels "1 ct TW" or "1 ctw" (which stands for 1-carat total weight) to mean "1 ct," thinking one stone weighs 1 carat. A ring with a "1 ct" top-quality diamond can be worth more than 10 times that of a ring with "1 ctw" of diamonds of the same quality.

When pricing diamonds, think in terms of the per-carat cost. To calculate the total cost of a diamond, use this equation:

**Total cost of a stone = carat weight × per-carat cost**

(opposite page) Early gem traders used carob beans as counterweights to balance their scales. *Alp Aksoy/ Shutterstock*

Diamonds can be divided into weight categories that have been determined by pricing guides. The weight categories for diamonds

weighing less than a carat may vary from one dealer to another. The weight categories below are based on those listed in the *Rapaport Diamond Report*, the most widely used diamond pricing guide by diamond dealers. As diamonds move up from one weight category to the next, their prices could increase anywhere from 5 to 50 percent.

| **Weight Categories for Diamonds** | | | | | |
|---|---|---|---|---|---|
| 0.01–0.03 ct | 0.04–0.07 ct | 0.08–0.14 ct | 0.15–0.17 ct | 0.18–0.22 ct | 0.23–0.29 ct |
| 0.30–0.39 ct | 0.40–0.49 ct | 0.50–0.69 ct | 0.70–0.89 ct | 0.90–0.99 ct | 1.00–1.49 ct |
| 1.50–1.99 ct | 2.00–2.99 ct | 3.00–3.99 ct | 4.00–4.99 ct | 5.00–5.99 ct | 10.00–10.99 ct |

There are a few exceptions to the rule that as diamond weight increases per-carat value increases. Because of high demand, diamonds weighing 5, 10 or 15 points could cost more per carat than larger odd-sized diamonds. Also, better quality 0.01-carat to 0.035-carat diamonds have at times cost more per carat than diamonds weighing 0.06 carats to 0.08 carats. This was the case from 1987 to 1989, when there was an unusually high demand for three pointers because of their use in tennis bracelets.

The complexity of diamond pricing can be discouraging, but you do not need to know the details of the system to shop for value. Just remember that carat weight as well as other factors can affect the per-carat value of diamonds and follow these two guidelines:

- Compare the per-carat cost instead of the total cost of a diamond. For example, if a diamond weighs 2 carats and costs a total of $10,000, it would be best to compare its per-carat cost of $5,000 to the per-carat price of other diamonds.
- When judging prices, compare diamonds of the same size, shape, quality and color.

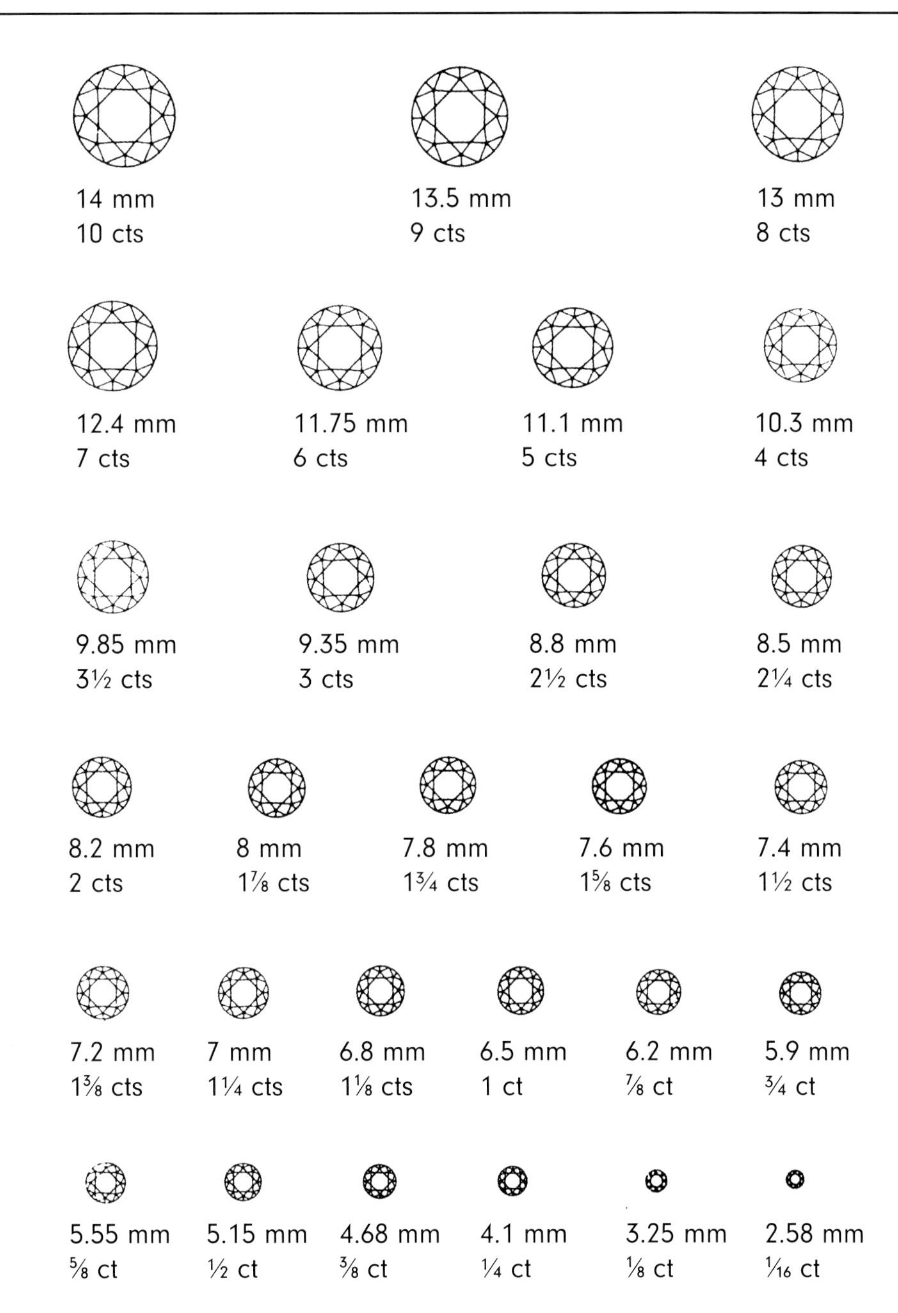


Approximate diameters and corresponding weights of well-cut round brilliant diamonds. (Note: 1 inch equals 25.4 millimeters.) *Diagram © GIA; reprinted with permission*

Diamond with high brilliance from The Gem Lab. *Photo by Paul Cassarino of the Gem Lab*

## Cut Quality

The term "diamond cut" is confusing because it can refer to both the shape and cutting style of a diamond (which we will discuss in the next section) and the quality of the cut, which is also called the "craftsmanship," "make," "cut quality" or "cut grade" of the diamond. The quality of the cut can be controlled by the cutter and is extremely important because it affects the beauty and brilliance of the diamond.

Brilliance is defined by the American Gem Society (AGS) as "brightness with positive contrast effects." Brightness is the actual and/or perceived amount of light returned by a diamond. A brilliant diamond has a pleasing, regular pattern of sharp, bright and dark areas.

When judging cut, you should assess the following:

- the proportions of the pavilion, crown, table and girdle;
- the finish, which is subdivided into polish and symmetry; and
- the optical attributes or light performance, i.e., brightness, fire, sparkle (scintillation) and contrast.

Cut quality assessment involves two fundamental considerations:

1. **Do you see brilliance all across the diamond when it is faceup?** Diamond brilliance should not be interrupted by large dark areas or white doughnut-shaped areas.
2. **Are you paying for excess weight that reduces the faceup size?**

These are separate factors. A diamond can have high brilliance yet have bulky proportions that make it look small for its weight when facing up.

You should evaluate diamond brilliance both with the naked eye and under magnification using a diffused light source such as a fluorescent lamp with a translucent white cover. Spotlights and bare light bulbs accentuate sparkle and fire, but they can also create dark shadows. Light that passes through a translucent bulb or material diffuses light more evenly. When selecting diamonds, however, examine them in a variety of lighting conditions and away from direct lighting. The best diamonds look good in all types of lighting.

A well-cut diamond with good brilliance. *Photo by Paul Cassarino of the Gem Lab*

An emerald-cut diamond with good brilliance. *Photo © Renée Newman*

A round diamond with a deep pavilion and dark center. *Photo © Renée Newman*

An emerald-cut diamond with a large bow tie. *Photo © Renée Newman*

A round diamond with a shallow pavilion and a white circle (fisheye). *Photo © Renée Newman*

Two views of a diamond with excellent brilliance, fire, polish and symmetry. Top-quality diamonds like this one display brilliance throughout and no see-through effect, even when tilted. *Photos by Michael Cowing, ACA Gem Lab*

Good diamonds display brilliance throughout the stone when faceup. They should not have dark or washed-out areas; you should not be able to see through the bottom of the stone. Fancy cuts such as the pear, marquise, oval or emerald cut may display a dark bow tie. The larger and darker the bow tie, the less desirable the stone. Most fancy-shaped diamonds have at least a slight bow tie, but when it is so pronounced that it is distracting, the bow tie lowers the value of the stone. Occasionally other dark patterns, such as a cross on an emerald cut, are visible.

Sometimes the faceup view of a round diamond displays a white doughnut-shaped area. In the diamond trade, this is called a fisheye. It is caused by the reflection of the girdle in a diamond with a pavilion that is too shallow. The thicker and more prominent the white ring, the poorer the cut. Besides looking bad, fisheye diamonds usually lack the brilliance of well-cut diamonds. On the other hand, if the pavilion of a round diamond is too deep, the center of the diamond will look dark.

Well-cut diamonds also display good brilliance throughout if they are tilted back and forth when viewed faceup. Faceted diamonds will reflect flashes of light when the stone, light source or observer moves. This effect is called scintillation. When viewed in spotlighting, well-cut diamonds also have a good display of rainbow colors called fire, which is technically referred to as dispersion.

In addition to having high brilliance and fire, the finest diamonds have excellent symmetry and polish. Their surface is blemish free, and their facets are well aligned and well shaped.

## Cutting Style and Shape

The shape of a diamond can play an important role in determining its price. For example, a 1-carat colorless round diamond might cost 15 to 30 percent more than a 1-carat square diamond of the same color grade and quality.

There are a variety of reasons why round and fancy-shaped (non-round) diamonds are priced differently. In weights above 0.25 carats, rounds usually cost more than other shapes because they are in higher demand. However, the percentage difference can vary depending on the size and quality of the stone, the supply and demand at the time of sale, and the dealer selling the stones. Since the pricing of the different shapes is complex, it is easiest for consumers to simply compare stones of the same shape when pricing diamonds.

The shape of a rough diamond crystal before it is cut also plays a role in pricing. When long diamond crystals are cut into ovals, pears or emerald cuts, they normally weigh more than if they had been cut into a round. This means the fancy shape's per-carat cost can be lower than that of a round and still bring the same amount of profit from the original rough diamond. Another advantage of buying fancy shapes is that they often appear larger faceup than round diamonds of the same weight.

If a square octahedron is cut into a princess cut, more weight will be retained from the rough than if it were cut into a round stone. This is a major reason why princess cuts and square emerald cuts generally sell for less per carat than rounds of the same weight and quality. Small rounds less than eight points are sometimes priced lower than small fancy shapes of the same weight and quality because of lower inventory costs and lower labor costs. The additional labor cost of small fancies is partially due to the specialized skills needed to cut them. It is less time consuming to cut, measure and select small rounds than small fancy shapes. In addition, small rounds sell more quickly, so less of a profit margin is needed to cover the cost of keeping them in inventory.

Fancy shapes may display a stronger faceup color than rounds cut from the same rough. This is one of the main reasons intense yellow round-cut diamonds cost more than intense yellow radiant-cut diamonds. The rough of the round diamond must be darker than that of the radiant to achieve the same intense yellow faceup color, partly

Diamond shapes and cutting styles.
*Image from KT Diamond Jewelers*

because cutters cannot play as much with the angles and shape of rounds to maximize the color. The stronger the color of the rough, the higher the price of the rough. The final price of the diamond is based to a large degree on the cost of the rough.

The prices of fancy shapes vary depending on what is most in style. In the 1980s, marquise diamonds were in high demand and could cost more than rounds of the same weight and quality. In 2020 there was a weaker demand for marquise shapes, so they generally sold for less per carat than oval-, pear- and cushion-shaped diamonds of the same weight and quality. However, demand for marquise shapes is increasing again. There has also been weak demand for princess cuts recently. One reason is that many divorcees who had received princess-cut engagement rings when they were especially popular were reselling them for cash. According to the *Rapaport Diamond Report*, ovals have been the best-selling fancy shape as the market has recently moved from squares to curves.

Cutting style can also have a slight effect on prices. Full-cut round brilliants sell for more than single cuts, which have fewer facets. New styles for cutting diamonds are continuously being developed. The majority of these nontraditional cuts are given brand names, and there can be up to a 10 percent premium on branded diamonds. Branded cutting styles usually cost more than generic cuts of the same size, shape and quality because of their higher production and advertising fees. Sometimes the price difference is due to the quality of the cutting. Often, one of the biggest advantages of branded fancy shapes is their consistent cut quality.

## CLARITY

Clarity is defined as the degree to which a stone is free from external marks, called blemishes, and internal features, called inclusions. Five factors of clarity grading are size, number, location, nature and relief (the contrast and color of the inclusions). The fewer, smaller and less noticeable the blemishes and inclusions, the better the clarity and the higher the price of the stone. Some might refer to blemishes and inclusions as imperfections or flaws. Gemologists usually prefer not to use these terms because of their negative connotations.

It can be to your advantage to buy a diamond with inclusions and blemishes. They can be proof that your diamond is untreated and of natural origin; they can act as a fingerprint to help protect you from having your diamond switched; they can lower the price of the diamond without affecting its beauty; and they can make you feel like your diamond is special. So if you are buying a diamond for personal enjoyment, do not worry about finding a flawless one; simply be concerned about inclusions that will make your diamond less attractive and less durable, such as large fractures.

There are various clarity grading systems, but the best-known system, which is used worldwide, is the one developed by GIA in 1953. Any knowledgeable diamond dealer or jeweler in the world will be able to understand you if you use the GIA system. Clarity grade descriptions of all the major systems assume a trained grader is working with 10 times fully corrected magnification and effective illumination. GIA defines its 11 clarity grades under these conditions as follows:

| **GIA CLARITY GRADES*** <br> *** For trained graders using 10 times magnification and proper lighting** | |
|---|---|
| **FL** | **Flawless**: no surface blemishes or inclusions |
| **IF** | **Internally flawless (loupe clean)**: no inclusions and only insignificant surface blemishes |
| **$VVS_1$ and $VVS_2$** | **Very, very slightly included**: minute inclusions ranging from extremely difficult ($VVS_1$) to very difficult ($VVS_2$) to see |
| **$VS_1$ and $VS_2$** | **Very slightly included**: minor inclusions ranging from difficult ($VS_1$) to somewhat easy ($VS_2$) to see |
| **$SI_1$ and $SI_2$** | **Slightly included**: noticeable inclusions that are easy ($SI_1$) or very easy ($SI_2$) to see |
| **$I_1$, $I_2$ and $I_3$** <br> In Europe: <br> **$P_1$, $P_2$ and $P_3$** | **Imperfect (piqué)**: obvious inclusions that are usually eye visible when viewed faceup. In $I_3$, distinctions are based on the combined effect on durability, transparency and brilliance |

A diamond with a GIA clarity grade of $I_1$ viewed at 10 times magnification under darkfield illumination. *Photo © Renée Newman*

The same $I_1$ diamond viewed at 10 times magnification under overhead lighting. *Photo © Renée Newman*

When professionals use microscopes to judge clarity, they normally examine the stones with a lighting setup called darkfield illumination. This uses diffused lighting from the sides against a dark background. (A frosted or shaded bulb provides diffused light, while a clear bulb does not.) Under this lighting, tiny inclusions and even dust particles will stand out in high relief. As a result, the clarity of the stone appears worse than it would under normal conditions.

Besides making inclusions and blemishes more prominent, darkfield illumination hides brilliance. To accurately assess the beauty, brilliance and transparency of a diamond, you should view it under overhead lighting (lighting above the stone), which is the way you normally view jewelry and gems. (Overhead lighting is reflected off the facets, whereas darkfield lighting is transmitted through the stone.) The quickest and easiest way to see a diamond's beauty magnified is with a good loupe at 10 times magnification.

If you ask a salesperson to show you a diamond under a microscope, it is unlikely that they will use its overhead lamp. Instead, they may only have you view the stone under darkfield illumination. Bear in mind, the inclusions will look more prominent than under overhead lights. To get a balanced perspective of the stone, also look at the diamond with a loupe and light above the stone, which is what diamond grading labs also do to determine the final grade.

## Transparency

GIA and the book *Gems* by Robert Webster define transparency as "the degree to which a gemstone transmits light without appreciable scattering." They list five categories of transparency, from most transparent to least, as follows:

- **Transparent:** Objects seen through the gemstone look clear and distinct.
- **Semitransparent:** Objects seen through the gemstone look slightly hazy or blurry.
- **Translucent:** The gemstone is cloudy and milky, like frosted glass.

- **Semitranslucent or semiopaque:** Only a small fraction of light can pass through the gemstone, mainly around the edges.
- **Opaque:** Virtually no light can pass through the gemstone.

Mineralogists use the term diaphaneity, but gemologists prefer the term transparency because it is easier for most people to understand.

Transparency can play a major role in determining the value and desirability of a gemstone. In most cases, the higher the transparency the more valuable the gemstone.

The famous 17th-century French gem merchant Jean-Baptiste Tavernier described diamonds with exceptional transparency and absence of color as "gems of the finest water" because such diamonds can look so colorless that they look like crystal clear water. The Gübelin Gem Lab offers a Finest Water Appendix to its diamond reports for large, very transparent and colorless Type IIa new diamonds (see chapter 1 for information about diamond types).

Many of the large historic diamonds of exceptional transparency are known to have come from mines near the diamond-trading center of Golconda in India. As a result, "Golconda" is another term occasionally used in the trade to describe an extremely transparent diamond.

The word "Golconda" is also used to describe highly transparent diamonds with other desirable attributes. For example, the Gübelin Gem Lab issues a Golconda Appendix together with its diamond grading reports for old diamonds that comply with a specific set of criteria, including D color, high transparency, Internally Flawless or improvable to IF, at least 5 carats in size, Type IIa and cut in an antique style. A mandatory step and a condition for getting a Golconda Appendix is that the lab must see the stone before it is repolished in order to ensure that it is not a new diamond with an antique cutting style.

The issuing of a Gübelin Golconda Appendix is based on quality criteria known to be typical of the famous Golconda diamonds and is not a determination of the geographic origin in the strict sense. In other words, a diamond with a Golconda Appendix is not necessarily from mines near the Golconda marketplace, and, conversely,

A diamond with high transparency. *Photo © Renée Newman*

A slightly cloudy semitransparent diamond. *Photo © Renée Newman*

A cloudy translucent diamond with a cloud in the center. *Photo © Renée Newman*

A 6-carat diamond that fulfils all the criteria to earn it a Gübelin Golconda Appendix. *Photo by Gübelin Gem Lab with kind permission of Lee Siegelson*

diamonds originating from the Golconda area are not necessarily of high quality. Type IIa D-color IF diamonds have been found in other countries besides India, and even in the U.S.

Clarity and transparency are interconnected, but they are different. If there is a cloudy spot in a transparent diamond, the cloud is a clarity feature. If the entire diamond is cloudy due to submicroscopic inclusions, then the cloudiness is a matter of transparency. If the diamond is very cloudy, this can affect the clarity grade. In most cases, however, clarity grades are not an indication of the degree of transparency of a diamond because diamond grading labs do not normally take into account subtle differences in transparency. Diamonds with VS clarity grades can be hazy and slightly cloudy, and diamonds with imperfect clarity grades can be transparent.

To check for transparency, examine the diamond at different angles and check to see if it is as clear as crystal glass or pure water. Faceup, the diamond should be brilliant, and there should be a strong contrast between the dark and bright areas. When judging transparency, make sure the diamond is clean, and examine it both with your naked eye and under 10 times magnification. You should also look at it under different lighting conditions (diffused fluorescent light, incandescent light, sunlight and away from light). Keep in mind that white objects or walls can reflect into the diamond, making it appear less transparent than it really is. It is helpful to have a highly transparent diamond sample for comparison. Nuances of transparency and haziness will be more easily detected, and your evaluation will be more accurate.

Some people mistakenly believe that you can determine a diamond's transparency by its degree of fluorescence and that diamonds with a medium to strong fluorescence will look milky or oily. Highly fluorescent diamonds can be perfectly transparent. The front cover of the Winter 1997 issue of *Gems & Gemology* showed a Harry Winston transparent diamond necklace that included some diamonds with medium to strong fluorescence, and Harry Winston would never have used a dull or low-grade diamond in his jewelry.

Sometimes consumers wonder why two diamonds with the same weight, shape, color, clarity and cut grades can have very different prices. One reason may be a distinct difference in their transparency.

Also, color, clarity and cut grades represent a range of qualities. A high $SI_2$-graded diamond can look much better than a diamond with a low $SI_2$ clarity grade.

In addition, not all sellers disclose if the color and clarity are natural or the result of treatments. These are discussed in the next section.

## Treatment Status

Unlike colored gems, most diamonds are untreated. However, that is changing; you can no longer assume that their color and clarity are natural. It is wise to ask if diamonds are untreated and accompanied by lab documents from reputable independent grading labs. Diamonds can undergo several treatments to improve their clarity, color, transparency and marketability.

### FRACTURE FILLING

This process improves clarity and transparency by filling extremely narrow cracks with a substance that makes them almost invisible. The filler used is a glass-like thin film, so the process does not add measurable weight to the stone. Even though you may not see them, the filled fractures are still present in the diamond. Another name for the diamond-filling process is glass infilling. A broader term is clarity enhancement. Clarity-enhanced (CE) diamonds are either fracture filled or laser drilled. According to the *GemGuide*'s market information, the discounts for fracture-filled diamonds can range from 20 to 40 percent. Be aware that direct heat, acid, repolishing or repeated cleaning procedures could remove or damage the fillers.

### LASER DRILLING

Laser drilling is another type of clarity enhancement. Its purpose is to get rid of dark inclusions. A focused laser beam drills a narrow hole into the dark area of the diamond. If the inclusion is not vaporized by the laser itself, then it is dissolved or bleached with acid. After the treatment, the hole looks like a white dot when the stone is faceup and like a thin white line when viewed from the side. Laser drilling is a permanent treatment. The dark spots will not reappear later on and

Laser-drilled holes in a diamond.
*Photo © Renée Newman*

lower the clarity grade of the stone. The discounts for laser-drilled diamonds can range from a few percent to 30 percent.

Sometimes the holes of laser-drilled diamonds are filled so that they will not be visible from the side. If the holes are filled, they can be as hard to spot as filled fractures. Such stones are considered to be both filled and drilled.

A 1.15-carat princess-cut diamond before irradiation and annealing. *Diamond and photo courtesy of Lotus Colors Inc.*

The same diamond after irradiation and annealing. *Diamond and photo courtesy of Lotus Colors Inc.*

### COATING

Diamonds are occasionally coated to improve their color grade. A fluoride coating such as that applied to eyeglass lenses can mask the diamond's yellow body color. Usually the coating is applied to the pavilion, but occasionally a thin coat is applied only at or near the girdle. Diamonds are also coated to change their color. Nail polish, enamel, silicon, metal oxide thin films and other substances have been used. Coatings on diamonds are not permanent and can be damaged during jewelry repair.

### IRRADIATION AND HEATING (ANNEALING)

Brownish and yellowish diamonds in the J to U color range and F to H color range are sometimes irradiated and heated (annealed) in an oxygen-free environment to temperatures of 850°F (450°C) and upward to produce a variety of colors. Irradiation gives diamonds a greenish-blue or blue color, and annealing them stabilizes the colors and creates new colors. F to H colors are used for light blue enhanced diamonds. The entire process is identified as "irradiated" on lab reports and in the rest of this chapter.

Yellowish diamonds that are irradiated can turn pure yellow or yellowish green. Brownish diamonds that are irradiated can become medium to dark blue or green, brownish yellow (golden yellow), brownish orange (orangish cognac), brownish red (reddish cognac), purple or black. The results can vary depending on the diamond.

The color of irradiated diamonds is basically stable, but some stones can change color if they come into contact with a jeweler's torch.

### LOW-PRESSURE HIGH-TEMPERATURE (LPHT) TREATMENT

Most black diamonds today have been created by heating heavily included diamonds at low pressures to temperatures above 2,372°F (1,300°C) to graphitize the fractures and turn them black. Unlike irradiated black diamonds, which usually have a bluish to greenish tint in transmitted light, heated diamonds are pure black. Besides being blacker, heated black diamonds cost less than irradiated blacks, but heated blacks are not as resistant to corrosive acids as irradiated black diamonds. Strong acids can give heated black diamonds a burnt, whitish look.

### HIGH-PRESSURE HIGH-TEMPERATURE (HPHT) TREATMENT

This process was first used to change the color of diamonds in the 1970s. Laboratories were able to produce yellow and green colors by heating diamonds to temperatures above 2,372°F (1,300°C) under extreme pressure. It was not until 1999 that the trade learned it was possible to turn inexpensive brown diamonds colorless using the same treatment process. In other words, D-, E- and F-color diamonds can be produced by HPHT treatment. Sometimes these diamonds are identified as "processed." If the color grade on a lab report is followed by an asterisk, check to see if there is a comment indicating that the diamond has been treated to change its color.

Colored HPHT diamonds were introduced to the market in 2000. Yellowish-green colors are the most common, but other colors, such as blue, pink and red, are also being produced. Consumers who cannot find natural stones in these colors now have the option of purchasing HPHT-treated counterparts that cost far less. HPHT-treated diamonds do not require any special care during cleaning or jewelry repair.

A big advantage to buying color-enhanced diamonds is the price. They are a fraction of the cost of untreated fancy color diamonds. As a result, more people can enjoy colored diamonds.

The effect of size, clarity, shape and cut quality on price is the same for color-enhanced diamonds as it is for natural color diamonds, but

Color-enhanced diamonds. *Diamonds and photo courtesy of Lotus Colors Inc.*

the way color affects price is different. According to Lotus Colors Inc., a company that specializes in color enhancement, light pink is the most expensive color for enhanced diamonds, followed by medium pink, purplish pink and light blue. Black is the least expensive color.

Sometimes HPHT-treated diamonds have inscriptions on the girdle indicating that they have been enhanced, but usually specialized skills and equipment are required to detect diamonds treated by the HPHT process. Your best insurance for determining if a diamond is of natural color is to get a report from a gem lab that is experienced in identifying irradiated and HPHT-treated diamonds.

## Creator (Natural or Lab Grown)

Diamonds created by nature are called natural diamonds or mined diamonds. Diamonds created in a laboratory or factory are identified by marketers as lab-grown, lab-created, created or simply lab diamonds. Gemologists and natural stone dealers often refer to them as synthetic, HPHT, CVD, HPHT-grown or CVD-grown diamonds (CVD is the abbreviation for chemical vapor deposition, a technique for growing diamonds). When the term synthetic diamond is used in gemological literature, it refers to lab-grown diamonds, not imitation diamonds. Lab-grown diamonds are not fakes; they are diamonds made by people instead of nature. Unlike cubic zirconia, which is an imitation diamond, lab-grown diamonds have the same chemical composition and crystal structure as mined diamonds.

As lab-grown diamonds have become more readily available, their prices have dropped significantly, with discounts of 35 percent or more. Pink and blue lab-grown diamonds are a fraction of the cost of natural diamonds. Chapter 7 explains how lab-grown diamonds are made and used.

## Diamond Grading Reports

The GIA Gem Trade Laboratory issued the first diamond grading reports in 1955. The color and clarity grades of the GIA grading report are based on a system they developed in 1953. Prior to that time, diamond color was described inconsistently in the trade, with terms such as River and Top Wesselton, or using multiple grades, such as A, AA and AAA.

Since GIA was the first to develop a diamond grading system that was taught to the trade and used on independent lab documents, GIA diamond reports are the best known in the industry and enjoy a worldwide reputation. The volume of diamonds graded at GIA is so high that it can take a long time to have a diamond graded there. As a result, other gem labs have emerged. Some are more respected and stricter than others, and some will deal with both trade members and non-trade members. One way to find out which labs are the most respected in your country is to ask jewelers which lab reports they would prefer to accompany a secondhand diamond if they were to buy one from you. Another way to find out about the reputation of a specific lab is to do an online search and read the comments on forums to compare them with other labs. The reputation of a lab is reflected in the asking prices for the diamonds that accompany their reports, with all other factors being the same. For example, a 1-carat round brilliant G-color $VS_1$ diamond with an excellent cut graded by a reputable lab can sell for 10 to 15 percent more than a diamond with the same specs graded by a lab known for assigning higher grades than merited.

Some people wonder why you need a lab report when appraisals also provide identification, treatment and quality information. The answer is that major, reputable laboratories have greater expertise, more sophisticated equipment and more opportunities to examine important gems than the average jeweler, dealer or appraiser. Many labs also conduct research that helps them spot new treatments and synthetic stones. Some appraisers will only appraise diamonds accompanied by lab reports, especially if it is a colored diamond, wherein natural versus treated color will have a dramatic effect on value. Another advantage of having a grading document from a major lab is that its documents usually carry more weight than appraisals if you ever want to resell a diamond.

In addition to a lab report, it is advisable to get a separate appraisal report for high-value diamonds after they are mounted in jewelry. Besides being useful for insurance purposes, an appraisal can help verify that the diamond matches the one described in the diamond report as well as provide information about accent stones and metal mountings. You can find a list of appraisal organizations and independent appraisers who are gemologists and who have completed formal education on appraisal procedures, ethics and law at ReneeNewman.com. Click on "Appraisers."

A true diamond grading report is not an appraisal because it does not indicate a value or price. It is simply an independent report that identifies and describes an unmounted diamond and indicates if it is untreated and of natural origin. The moment a report includes a price, it becomes an appraisal.

## Non-Quality Factors that Can Affect Prices

Diamond grading reports usually indicate the degree of fluorescence of a diamond. This is considered an identification feature, not a grading factor. Strong blue longwave fluorescence helps prove that a diamond is natural and not lab grown. However, diamonds with strong to very strong fluorescence are often discounted because some jewelers have the mistaken belief that if a diamond has strong fluorescence it will look cloudy. Diamonds with strong fluorescence can have excellent transparency. As established earlier, it is best to determine diamond transparency by actually looking at the stone and comparing it to other diamonds rather than by checking its fluorescence.

Alrosa, a Russian group of diamond mining companies, is using diamond fluorescence as a positive selling feature. In November 2020, it launched a new brand called Luminous Diamonds that aims to attract jewelry buyers through glamorous designs set exclusively with rare fluorescent diamonds. The brand's marketing links the idea of fluorescence, a diamond's "inner glow," to a woman's "inner light."

Many other factors play a role in diamond pricing. These include demand, currency fluctuations, forms of payment, a buyer's credit

Diamonds photographed in typical daylight-balanced fluorescent lighting. *Photo by Michael Cowing, ACA Gem Lab*

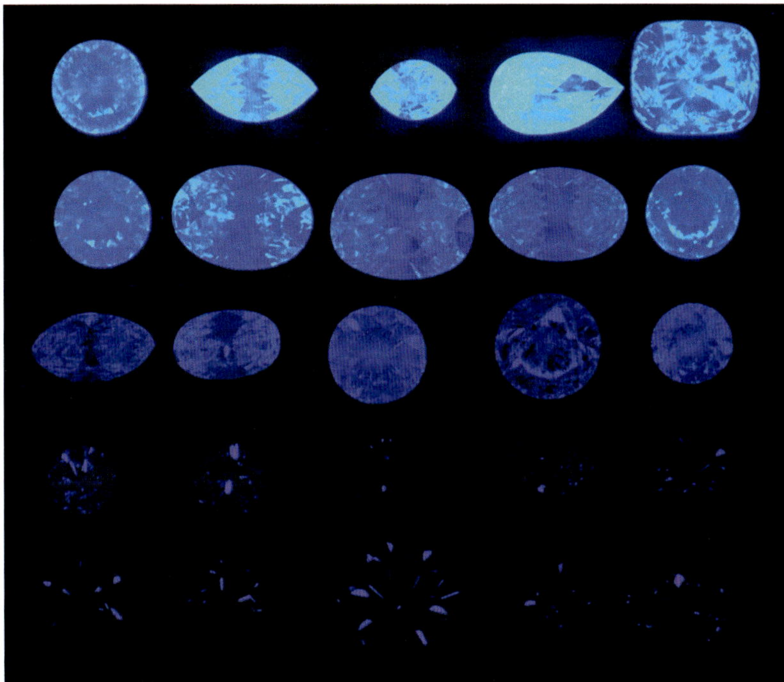

The same diamonds under a longwave UV lamp showing five rows of blue fluorescent strengths ranging from very strong, at the top, to strong, medium, faint and none, at the bottom. *Photo by Michael Cowing, ACA Gem Lab*

rating, quantity of diamonds purchased, time of sale, competitors' prices, a seller's need for money and a customer's eagerness to buy. The more valuable a diamond is, the harder it is to predict the price. Nobody was able to predict that the Hancock Red diamond discussed at the beginning of this chapter would sell for $880,000. Even though its red color was very rare, it weighed less than 1 carat and had an imperfect clarity grade. Its presale estimate was $100,000 to $150,000, yet it sold for eight times its expected price. Ultimately its value was determined by the price the buyer was willing to pay.

# 7 Laboratory-Grown Diamonds

In 1772 Antoine Lavoisier (1743–1794) and other chemists made a surprising discovery about diamond. Lavoisier and his colleagues enclosed a diamond in a glass jar and used a giant magnifying glass to focus the Sun's rays on it, causing it to burn and disappear. Despite the fact that the diamond had vanished, Lavoisier noticed that the overall weight of the jar did not change. He also noticed that diamond, when burned, produced the same gas as charcoal — carbon dioxide. From that, he realized that diamond and charcoal are different forms of the same element, which he named carbon. This discovery encouraged other scientists to try to work out a way to create diamonds from carbon.

A portrait of chemist Henri Moissan. *Science History Images/Alamy Stock Photo*

In 1892 French chemist Henri Moissan (1852–1907) theorized he could create diamonds by crystallizing carbon under pressure from molten iron in an electric arc furnace he had developed. The furnace reached temperatures up to 6,332°F (3,500°C), but the resulting material was not diamond. Moissan became more famous for his discovery of moissanite, which is also known as silicon carbide. He uncovered the rare natural form of it in 1893 in a meteorite that had landed in the Diablo Crater in Arizona. Today most moissanite is laboratory grown, and it is used in jewelry as a diamond substitute.

As you will see, it was not until the early 1950s that scientists finally achieved their dream of creating diamonds.

## Terminology for Laboratory-Grown Diamonds

Diamonds created in laboratories were first called synthetic diamonds, and many scientists and natural diamond dealers still use that term. However, to many consumers, the term synthetic means fake, so people who sell these diamonds prefer to identify them with other names, such as laboratory grown, lab grown, lab created, created or LGDs. Lab-grown diamonds have the same crystal structure and chemical formula as natural diamonds — carbon. Stones like cubic zirconia and moissanite have a different chemical composition and are called imitations, simulants or fakes.

Lab-grown diamonds can also be identified by their growth method. Diamonds grown under high pressures and high temperatures are called HPHT-grown diamonds or, simply, HPHT diamonds. Those that are produced using chemical vapor deposition techniques are called CVD-grown diamonds or CVD diamonds.

## HPHT-Grown Versus CVD-Grown Diamonds

HPHT diamonds are grown by placing minute synthetic diamond seed crystals and a carbon source (usually low-quality synthetic diamond powder) in a special pressure chamber with a molten metal flux, such as iron, nickel or cobalt (a flux is a material that dissolves other materials when melted). After applying extremely high pressure and high temperatures under controlled conditions, the carbon is dissolved in the molten flux. The carbon atoms migrate to the diamond seed crystals and bond to them, which grows the diamond. Within a few days or weeks, new HPHT-grown diamond crystals suitable for faceting are formed.

CVD-grown diamonds are produced at high temperatures and low pressure in a vacuum chamber with carbon-containing gases and synthetic diamond seed plates. The gas molecules are broken apart during decomposition in a microwave field, and carbon atoms are

## Growth Process of a CVD Lab-Grown Diamond

**Starts with a thin slice of diamond seed**

**Carbon melts and forms into a diamond around the seed**

**The carbon is removed from the diamond**

**The rough stone is cut and polished into a diamond**

*Image © Smiling Rocks Inc.*

deposited on the "seeds," causing them to grow into synthetic diamond crystals, which are afterward cut and polished into faceted stones. Today the CVD process is used to produce high-color (or fancy color) and high-clarity Type II diamonds up to several carats in weight.

By law, sellers must disclose to buyers whether a diamond is lab-grown. Unfortunately, there are unscrupulous sellers who try to pass synthetic diamonds off as natural, so it is important to find ways to recognize them. Various synthetic diamond detectors have been developed to spot man-made diamonds, and the prices range from $300 to many thousands. The safest way to determine whether a diamond is natural or synthetic is to send it to a gem laboratory that has the necessary high-tech equipment. Some appraisers will refuse to appraise diamonds if they do not have an accompanying lab report from a respected lab. Nevertheless, many but not all lab-grown diamonds can be detected with the basic methods below:

Additional information on the characteristics and identification of HPHT and CVD diamonds is available in *Laboratory-Grown Diamonds*, third edition (2020) by Branko Deljanin and Dusan Simic, *Gems & Gemology in Review: Synthetic Diamonds* (2008) by James Shigley, *Diamond Handbook*, third edition (2018) by Renée Newman and select issues of *Gems & Gemology* (Fall 2016, Fall 2017 and Summer 2018).

- **Magnification of inclusions, color zoning and growth patterns:** Natural diamonds may have included minerals such as garnet, diopside and diamond, which are not present in man-made diamonds. Gray or black metallic flux inclusions may be present in HPHT-grown diamonds, whereas the occasional dark inclusions in CVD diamonds are graphitic rather than metallic. Color zoning (uneven distribution of color) may be

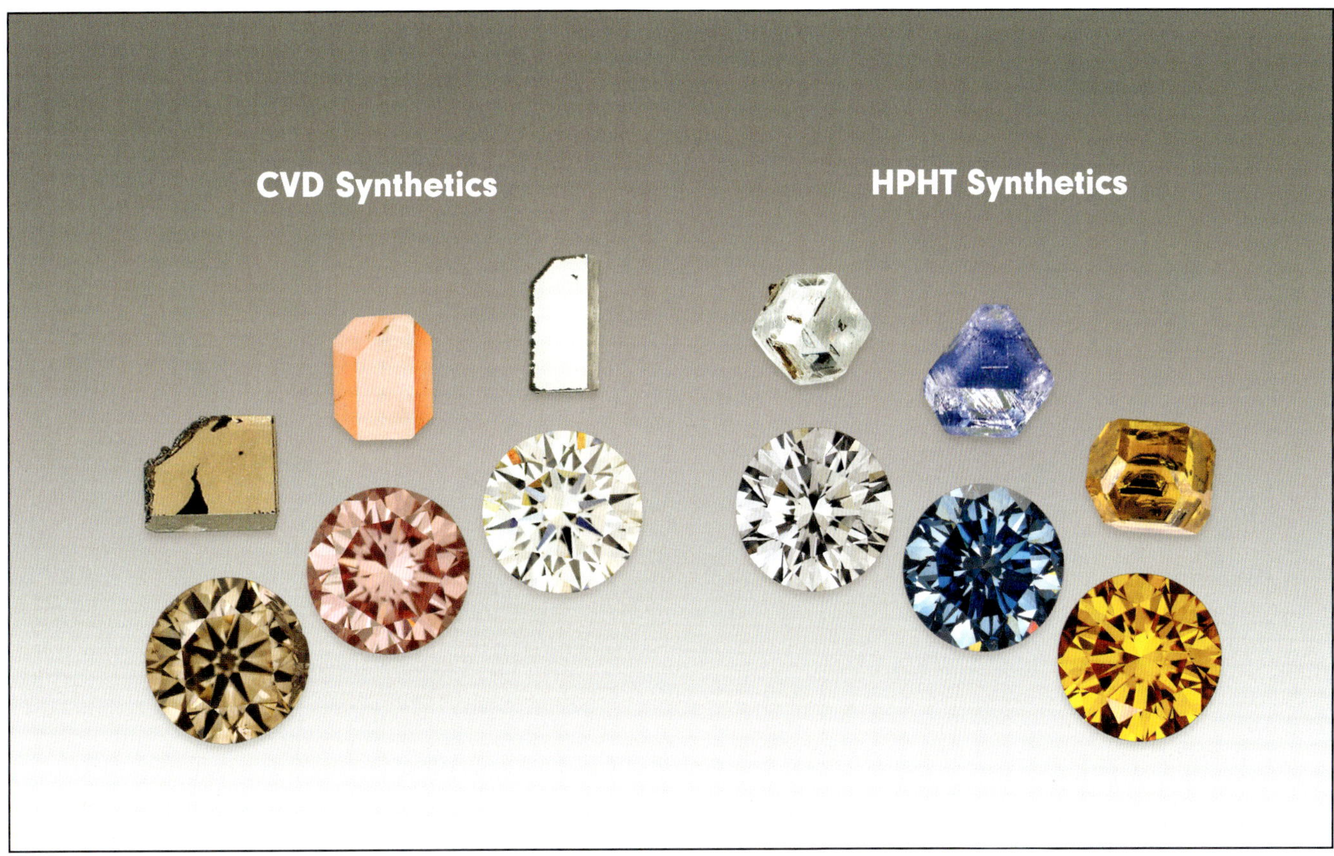


CVD- and HPHT-grown synthetic diamonds are made in a variety of colors, depending on the growth conditions and post-growth treatments. While their final faceted appearances may be similar, the crystals produced from the two techniques are usually quite different. *Image © GIA; reprinted with permission*

more pronounced in HPHT-grown diamonds than in natural stones. CVD-grown diamonds, on the other hand, are usually evenly colored. Since synthetic diamonds grow differently than natural diamonds, they have different growth patterns. These patterns are most noticeable under ultraviolet lighting.

- **Magnetism using a rare earth magnet:** HPHT diamonds sometimes have metallic inclusions that allow them to be attracted to rare earth magnets.
- **Fluorescence colors and strengths under longwave and shortwave UV:** Natural diamonds usually fluoresce stronger under longwave UV rays than under shortwave UV rays, whereas most synthetics fluoresce stronger under short-wave rays. If a yellow, colorless or near colorless diamond has strong or very strong blue longwave UV fluorescence, it is a natural diamond.

Even though the above methods can help determine whether a diamond is natural or synthetic, advanced spectroscopic instruments are generally required to confirm that a diamond is natural.

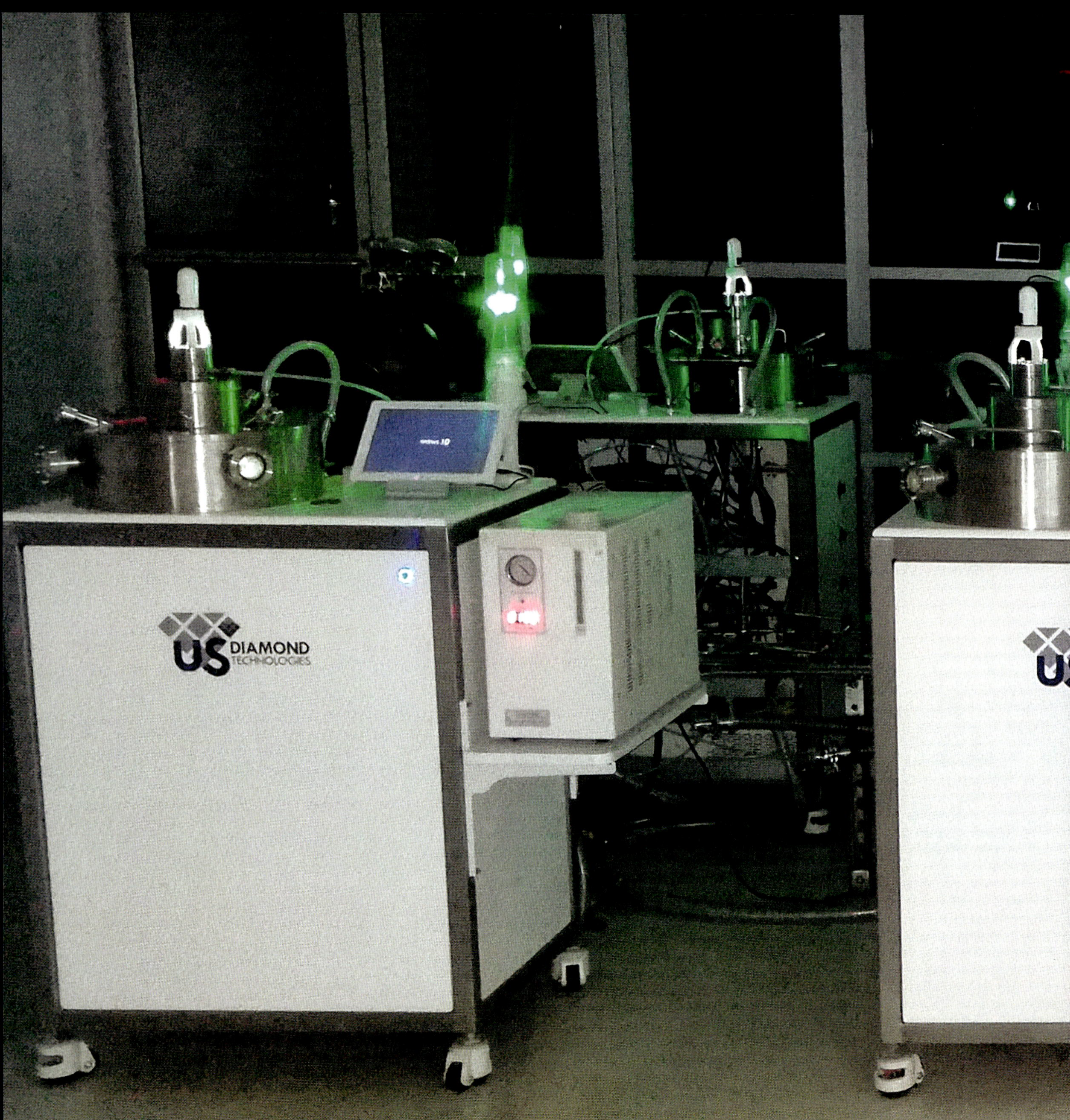
US
DIAMOND
TECHNOLOGIES

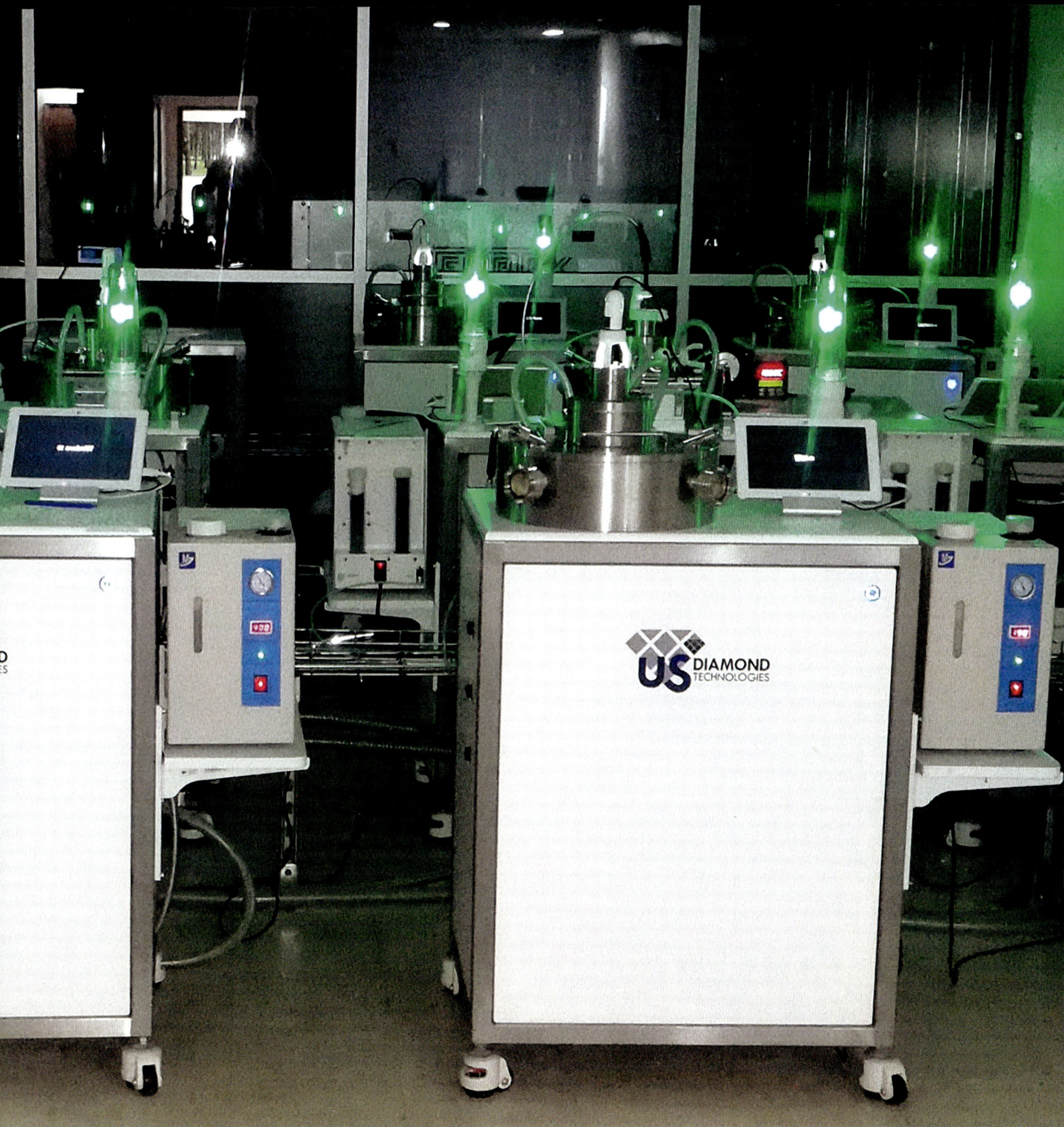


CVD diamond growers made by US Diamond Technologies, a producer of microwave plasma CVD reactors in New Jersey. *Image from* Laboratory-Grown Diamonds *(2020), courtesy of authors Branko Deljanin and Dusan Simic*

## Time Line of Lab-Grown Diamond Development

On December 16, 1954, Dr. Tracy Hall, a physical chemist at General Electric (GE), became the first person to create a synthetic diamond by a reproducible, verifiable and witnessed process. He used a press of his own design. However, he was not the first person to create a diamond, and the stone he produced was a small particle of HPHT diamond for industrial use.

Over two decades earlier, in 1929, Dr. J. Willard Hershey of McPherson College in Kansas had repeated Moissan's diamond synthesis experiment and created two synthetic diamonds. He continued his experiments in the 1930s, which he wrote about in *The Book of Diamonds: Their Curious Lore, Properties, Tests and Synthetic Manufacture* (1940). Hershey claimed to have made more than 50 diamonds ranging in size from 1 to 2 millimeters by 1 millimeter.

In 1952 William G. Eversole of the Union Carbide Corporation recorded the first successful attempt at producing a CVD diamond, and in 1953 scientists in Sweden, at Allmänna Svenska Elektriska Aktiebolaget (ASEA), grew HPHT diamonds but kept the discovery secret until much later.

Here is a time line of the many important developments that have occurred in the history of growing synthetic diamonds:

---

### 1950s

**1952:** William G. Eversole creates a tiny CVD diamond at the Union Carbide Corporation.

**1953:** The Swedish electrical company ASEA reportedly produces the first HPHT diamonds, but it keeps the discovery a secret until the 1980s.

**1954:** Tracy Hall at GE produces a small particle of HPHT diamond for industrial use with a press of his own design. Thanks in part to Hall's work, GE was a dominant player in industrial diamond development for many years. Hall received a $10 savings bond from GE for his invention, and he was awarded the American Chemical Society Award for Creative Invention.

**1956:** The Russian scientist Boris Spitsyn discovers how to grow CVD polycrystalline diamond thin films on non-diamond substances.

---

1960s

**1960s:** GE and De Beers Industrial Distributors Limited produce HPHT industrial diamonds.

---

1970s

**1970:** GE creates the first experimental gem-quality HPHT diamond.

**1970:** Sumitomo Electric Industries of Japan begins developing HPHT diamonds after noticing that diamond could be used as a die material in place of cemented carbide, which was in use at the time.

**1971:** GE creates several gem-quality synthetic diamonds and sends them to Robert Crowningshield at the GIA Laboratory for examination. The diamonds weigh between 0.26 and 0.30 carats and range from F to J color. GE also creates several fancy color yellows and blues that GIA also analyzes.

**1976:** Commercial production of the diamond imitation cubic zirconia begins after Russian scientists perfect a technique for producing the mineral while attempting to grow diamonds.

---

1980s

**1982:** Sumitomo Electric synthesizes a 1.20-carat single crystal. This diamond is registered in the 1984 *Guinness Book of World Records* as the world's largest synthetic diamond. Subsequently Sumitomo Electric becomes the first mass-producer of approximately 1-carat synthetic single-crystal diamonds. They are yellow due to impurities, and their size is 5 millimeters or less.

**Mid-1980s:** De Beers, Sumitomo Electric and Russian firms all grow colorless HPHT diamonds larger than 1 carat.

**Late 1980s:** De Beers Industrial Distributors Limited starts production of CVD polycrystalline industrial diamonds and research into single-crystal CVD diamonds.

---

## 1990s

**1990s:** Russian firms produce the first HPHT diamonds for use in jewelry.

**1995:** GE patents the production of transparent polycrystalline diamond films.

**1995:** Ultimate Created Diamonds begins offering intense yellow HPHT diamonds to the jewelry trade.

---

## 2000s

**2000:** Sumitomo Electric succeeds in synthesizing a high-purity, clear and colorless 8-carat diamond crystal that measures 0.4 inches (1 centimeter) in diameter.

**2002:** Gemesis begins producing commercial quantities of HPHT diamonds for jewelry.

**2002:** De Beers changes the name of Industrial Distributors Limited to Element 6 (in reference to carbon, which is the sixth element of the periodic table). Today Element 6 is one of the largest synthetic diamond companies in the world and is the diamond provider for De Beers's Lightbox jewelry brand.

**2005:** Researchers at the Carnegie Institution for Science's Geophysical Laboratory produce 10-carat, half-inch (1.27-centimeter) single-crystal untreated CVD diamonds at rapid growth rates (100 micrometers per hour).

**2006:** Advanced Optical Technologies Corporation (Canada) starts producing HPHT-grown and irradiated and annealed blue, yellow, colorless and pink diamonds up to 1.50 carats.

Research scientist and gemologist Branko Deljanin inspects a 10.02-carat New Diamond Technology diamond at the Mediterranean Gemmological and Jewellery Conference in 2015. *Photo by John Chapman*

**2008:** WD Lab Grown Diamonds is founded and becomes the exclusive licensee of the single-crystal CVD diamond growth technology developed by the Carnegie Institution for Science.

---

## 2010s

**2014:** New Diamond Technology begins producing colorless HPHT diamonds up to 5.11 carats.

**2015:** New Diamond Technology makes the first colorless faceted lab-grown stone surpassing 10 carats — an HPHT-grown 10.02-carat E-color $VS_1$ diamond.

**2016:** New Diamond Technology makes a 10.07-carat emerald-cut Fancy Deep blue diamond. This remains the world's largest Fancy Deep blue lab-grown diamond produced to date.

**2017:** Companies in Zhengzhou, China, are each producing more than 1,000 carats of near-colorless HPHT diamonds per day. After faceting, most of the small synthetic diamonds weigh between 0.005 and 0.03 carats, with color ranging from D to N and widely variable clarities.

The largest yellow HPHT faceted diamond to date (20.22 carats). *Grown by New Diamond Technology and distributed by Meylor Global LLC. Photo courtesy of Meylor Global LLC*

**2018:** De Beers starts selling lab-grown diamond jewelry under the brand name Lightbox.

**2018:** New Diamond Technology grows a 103.50-carat HPHT diamond, which was the largest-known uncut lab-grown diamond until 2020.

**2018:** The company Unique Lab Grown Diamond submits a 5.01-carat $SI_1$ Fancy Intense pinkish-orange CVD diamond to GIA. The combination of size, color and clarity makes it the most remarkable CVD diamond tested by GIA to that date.

**2018:** WD Lab Grown Diamonds grows the largest CVD round brilliant-cut diamond to date — a 9.04-carat CVD as-grown diamond with no HPHT post-growth treatment.

**2018:** New Diamond Technology grows a yellow HPHT single-crystal diamond weighing 55.94 carats. It is cut into a cushion-shaped $VS_2$ clarity diamond weighing 20.22 carats. This cut diamond currently holds the world record as the largest faceted yellow HPHT diamond.

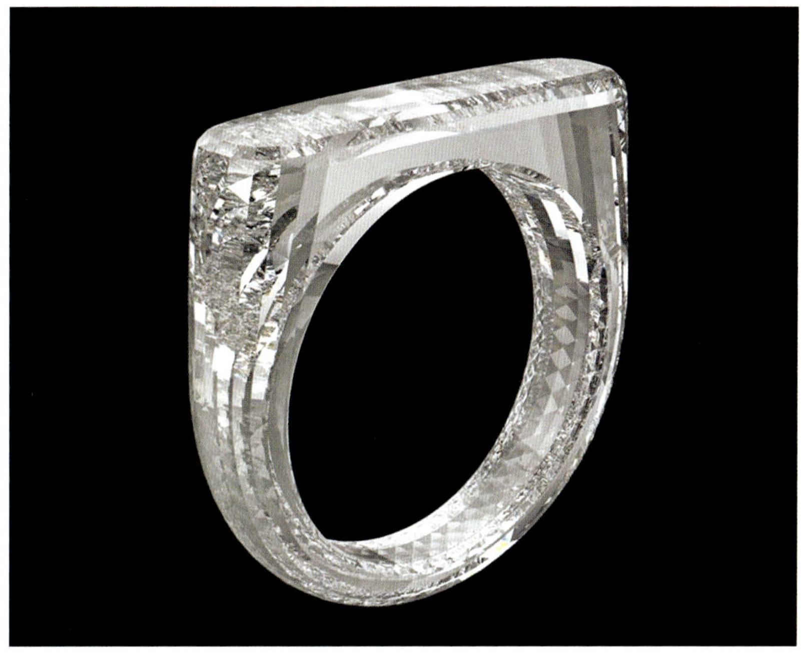

The first lab-grown all-diamond ring. Jonny Ive and Marc Newson joined up with gemstone manufacturer Diamond Foundry to create this ring for a (RED) charity auction through Sotheby's in 2018. The ring raised a total of $461,250. *Ring and photo by Diamond Foundry*

**2018:** The Diamond Foundry produces the first lab-grown all-diamond ring. It is designed by Apple's former chief design officer Jony Ive and industrial designer Marc Newson.

**2019:** Swedish engineering company Sandvik creates the first 3-D-printed diamond composite from a slurry of diamond powder and polymer using a method called stereolithography. The diamond composite is opaque, but it can be used in applications and shapes never before considered possible.

---

## 2020s

**2020:** Meylor Global LLC and Dr. Andrey Katrusha create the largest lab-grown uncut diamond, weighing 109.81 carats, in Kyiv, Ukraine. It is verified on August 19, 2020, by *Guinness World Records*.

**2020:** Lightbox, a De Beers–owned company that produces lab-grown diamond jewelry, officially opens its $94-million manufacturing facility in Gresham, Oregon, in October. It plans to produce approximately 200,000 carats of lab-grown diamonds annually.

Inside the Lightbox facility for growing CVD diamonds in Gresham, Oregon. *Photo © Lightbox Jewelry*

## Lab-Grown Diamond Jewelry

When lab-grown diamonds were first marketed for jewelry in the 1990s, they were sold as loose diamonds for jewelers to mount in jewelry, and their prices were not much lower than those of natural diamonds due to their rarity. However, as lab-grown diamonds became more widely available, their prices decreased. By 2015 they were selling for about 25 to 30 percent less than natural diamonds of the same weight, shape, color and clarity.

A significant drop in price occurred in 2018 when De Beers started selling lab-grown diamond jewelry under the brand name Lightbox. Unlike companies that price their lab-grown diamonds on the basis of their quality factors and weight, the prices of Lightbox diamonds set in jewelry are based only on their weight. All 1-carat Lightbox diamonds, for example, are the same price no matter what their color or shape. To compete with Lightbox and other new companies entering the market, diamond growers had to further lower their prices.

The main advantage of buying lab-grown diamonds is that they cost less than natural diamonds but have the same resistance to scratching, abrasion and chemicals. The biggest price difference is with colored lab-grown diamonds, which can be a small fraction of the cost of natural fancy color diamonds.

Another benefit of man-made diamonds is that they are easier to match in jewelry than natural diamonds because they can be produced in unlimited quantities under controlled conditions. As a result, they do not have a resale cash value like natural diamonds do. The supply of mined diamonds is limited, each diamond is unique, and they are formed in the interior of the Earth over millions of years.

The first lab-grown diamond jewelry that Lightbox and other companies offered consisted of basic solitaire rings, pendants and stud earrings. These have evolved into multi-stone pieces with innovative designs. In addition, the selection of lab-grown diamond shapes, sizes and colors has greatly increased. For example, lab-grown eternity bands are no longer limited to round brilliants; they may also be set with lab-grown diamonds of other shapes.

Eternity bands set with cushion-, oval- and emerald-cut lab-grown diamonds. *Rings and photos © Smiling Rocks Inc.*

A 2.75-carat (total weight) lab-grown blue and white diamond ring. *Ring and photo courtesy of New World Diamonds*

An ear-climber (or ear arc) lab-grown diamond earring. *Earring and photo courtesy of Vrai*

A pair of lab-grown diamond rings from Smiling Rocks. *Rings and photos © Smiling Rocks Inc.*

A selection of lab-grown diamond jewelry from De Beers's company Lightbox. *Photos © Lightbox Jewelry*

A reversible lab-grown diamond ring that can display either a red or blue lab-grown diamond center stone. *Ring and photos courtesy of New World Diamonds*

# 8

# Diamond's Remarkable Benefits

Arguably no other material offers as many benefits as diamond. That is because its unique properties make it superior to other competing materials. Before the 1950s low-grade diamond rough, grit and powder were used for machinery and cutting, drilling and grinding tools because of diamond's extreme hardness. After scientists discovered how to synthesize diamonds in the 1950s, natural rough was no longer the sole source of industrial diamond material, and its uses began to expand further. The development of chemical vapor deposition (CVD) diamonds and transparent polycrystalline diamond films have played major roles in the expansion of diamond applications.

Synthetic diamond grit. *Image used with permission of Element Six Group, Copyright © Element Six Group, 2020*

Sandvik 3-D-printed diamond composite. *Photo courtesy of Sandvik Additive Manufacturing*

A further development occurred in 2019 when Sandvik created the first 3-D-printed diamond composite and developed a proprietary post-production process that gives it extremely high hardness and exceptional heat conductivity. The new process means that this superhard material can now be 3-D printed in highly complex shapes, revolutionizing the way diamond material is used. Even though diamond composite is not transparent and does not sparkle, it is ideal for a wide range of industrial uses.

The benefits of diamond extend far beyond its use in creating better-functioning and longer-lasting machinery and tools. This chapter describes the amazing features of diamond and explains how they benefit us.

## Superior Hardness

Diamond's carbon atoms have a strong interlocking atomic structure and are exceptionally close together. As a result, diamond is the hardest natural substance, making it more resistant to scratching, abrasion, wear and distortion from pressure than any other material. Diamond coatings are now used to protect lenses and cell phone screens from scratching. In 2017 Rolls Royce created a paint made from 1,000 crushed diamonds for a custom-made version of its Ghost luxury sedan. The paint, named Diamond Stardust, is smooth, and the diamonds are undetectable to the touch. It may not be long before other car manufacturers are using diamond-based paints and enamels to reduce scratching.

Polycrystalline diamond compact drill bits like this one have wide applications in mining and oil and gas drilling. *ribeiroantonio/Shutterstock*

Diamond materials are invaluable in the agricultural, construction, automotive and manufacturing industries, which all need machinery that can cut, grind and drill. The advantage of diamond is that you can work with it at faster speeds, for longer amounts of time and with more predictable tool wear rates compared to other materials. It significantly increases the lifetime of instruments and equipment, making it more cost-effective from a long-term perspective.

A relatively new but rapidly growing application for industrial diamonds is in cutting wire for saws. Steel wire is coated in a resin that is then impregnated with diamonds ranging in size from powder, in the case of a very thin wire, to more than 1-millimeter in diameter, in the case of thick industrial cable. Unlike traditional saws, which generally cut in only one direction, diamond wire can cut rounded edges and other complex shapes. Diamond wire can also be used for very fine cuts, for example, to manufacture surgical equipment and computer parts. Currently the applications for diamond wire seem to be growing faster than production can supply.

Diamond cutting wire slicing through granite. *HSBortecin/Shutterstock*

Diamond speaker domes have been shown to produce outstanding sound quality. In addition to diamond being the hardest natural substance, it is also the stiffest. Vibration inside a speaker is the primary reason that sound quality degrades between the speaker and your ears. Diamond can be accelerated very rapidly without any physical deformation in shape, which is why it produces such high-fidelity sound. Researchers are using CVD diamonds to build speaker domes that are lighter than their aluminum counterparts and as stiff as any natural substance.

Ball bearings are found in many devices we use every day, such as cars and hard drives. They are so important to the functioning of vehicles that, in World War II, the Allied forces strategically attacked a German ball bearing manufacturer to cripple the German army. Ball bearings take a lot of punishment — such as high heat and friction — while in use. The hardness, resistance to abrasion and thermal conductivity of diamond makes it a perfect choice for ball bearings. Testing has shown that diamond bearings can last as much as eight times longer than tungsten carbide ones. The requirement to shape diamonds into perfect spheres is currently an impediment to wide-scale commercial use of diamond ball bearings, but that could change now that diamond composite can be 3-D printed into any shape.

## Nontoxic and Biocompatible

Diamond is safe to use on and in humans because it is nontoxic. This allows doctors and dentists to use synthetic diamond as well as diamond coatings and particles in surgical drilling and cutting tools, implants, artificial body parts and chemotherapy patches. Thanks to diamond coatings, hip and knee replacements can last longer than they would otherwise.

## Resistant to Chemicals and Radiation

Diamond cannot be damaged by chemicals. Its chemical resistance and superior hardness not only make it the perfect gemstone for everyday wear, they also make it an ideal material for corrosion-resistant coatings and use in machinery, space exploration and scientific instruments.

Diamond's resistance to radiation also makes it an ideal material for space exploration, defense programs and medical technology. Single-crystal HPHT and CVD diamonds are used in radiation detectors and monitors and in radiotherapy.

## Excellent Heat Conductor

Diamond's thermal conductivity is higher than that of any other solid material and five times higher than that of copper, which is why diamonds feel cool to the touch. This property makes diamond important as heat sinks (i.e., cooling agents) in lasers, electronics and other industrial uses because it protects silicon and other semiconducting materials from overheating. This also allows instruments that measure heat conductivity to easily distinguish diamonds from glass and cubic zirconia.

When electronic devices overheat, system performance and service life can quickly degrade. CVD-diamond heat spreaders can more than double overall system productivity. Companies can run at higher power levels without increasing junction-operating temperature by using CVD-diamond components.

## Excellent Electrical Insulator and Semiconductor

Non-blue diamond strongly resists electric currents, so it is very effective at blocking the flow of currents. Blue diamond, however, gets its color from boron impurities, which make it a semiconductor. Since diamond can be both an electrical insulator and a good conductor of heat, it is invaluable for the manufacture of appliances, cell phones and computers.

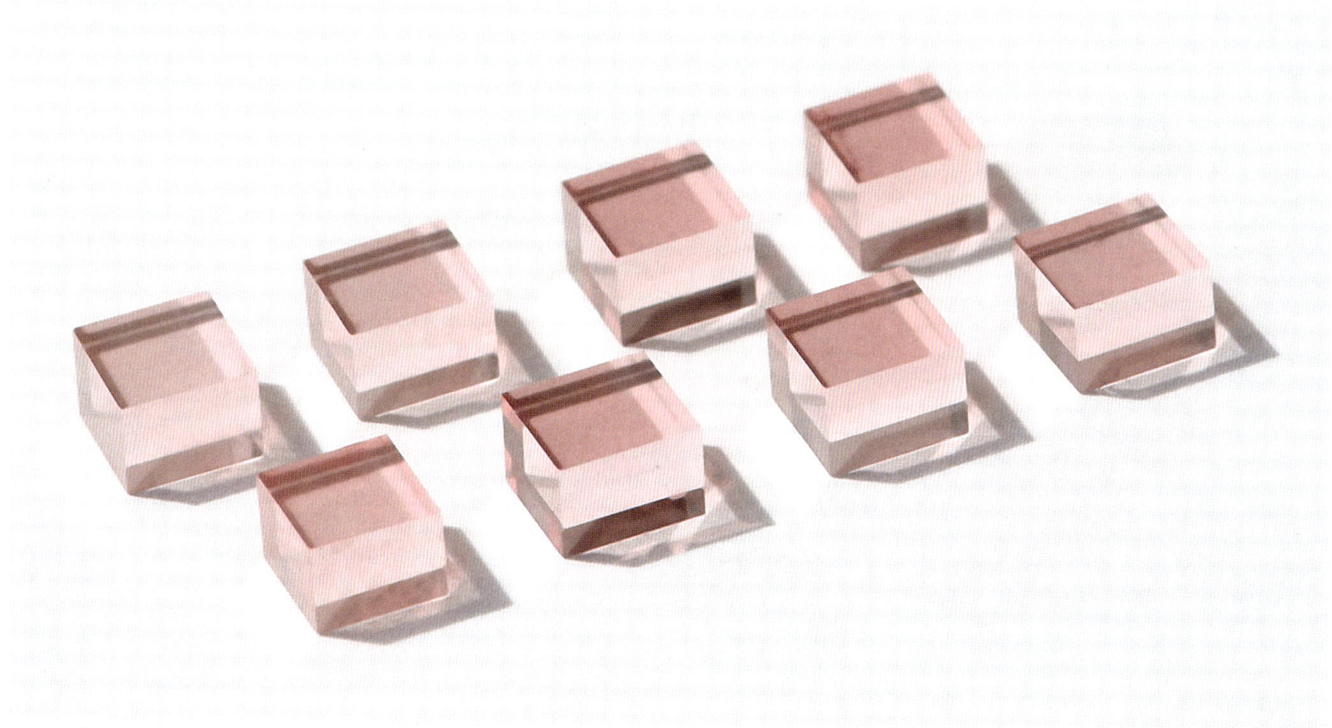

Single-crystal electronic-grade CVD-diamond material. *Image used with permission of Element Six Group, Copyright © Element Six Group, 2020*

PJP3
PD7
Y1
U1
ST3243EC
7B348
100
51
50

There are circumstances when synthetic diamond is intentionally doped with boron so that it can conduct electricity and be used to create an electrode-based system for water purification. Contaminated industrial wastewater can now be purified and disinfected without chemicals thanks to boron-doped lab diamond electrodes.

## Resistant to High Temperatures

Diamond's resistance to high temperatures permits jewelers to steam clean and repair diamond jewelry using a torch — unlike with many gems, which must be removed during repairs. An even more important benefit is that diamond's heat resistance, combined with its semiconductor properties, makes it the best material for electronics.

Compared to silicon, diamond can tolerate higher voltages before breaking down; it can run five times hotter without degrading in performance, and it is more easily cooled since it has 22 times the heat transfer efficiency of silicon. Semiconductor devices with diamond material are already available, and they deliver one million times more electrical current than silicon.

Adam Khan, founder and CEO of AKHAN Semiconductor, reported on WIRED.com that diamond-based semiconductors not only increase power density but also create faster, lighter and simpler devices. They are more environmentally friendly than silicon and improve thermal performance within a device. Diamond semiconductors can also help better manage battery life and battery systems for a wide variety of devices, including cell phones, cameras and vehicles. Because diamond technology shrinks the size and energy needed for a semiconductor, it paves the way for smaller personal electronics.

As for defense technology, diamond delivers greater range, reliability and performance in both normal and extreme/hazardous operating environments. Consequently, the diamond materials market for semiconductors can easily eclipse that of silicon carbide.

(opposite page) Semiconductors are used in everything from smartphones to washing machines to LED light bulbs. Many view diamond semiconductors as the future of the electronics industry. *Ksolstudios/Shutterstock*

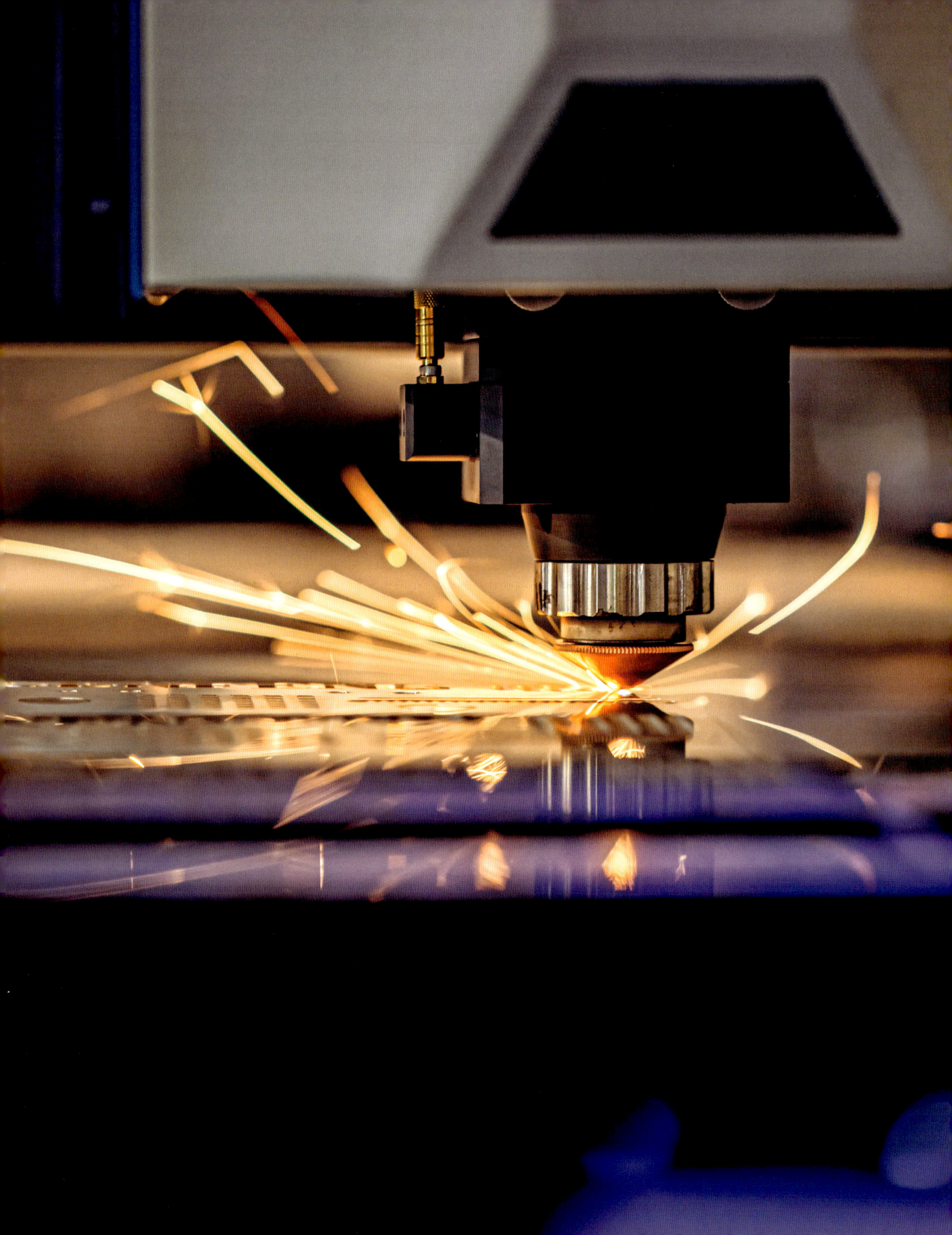

## Exceptional Transparency

Pure and defect-free diamond lets a wide spectrum of light, from ultraviolet to infrared, pass through it — unlike much glass, which is only transparent to visible light. Diamond's outstanding transparency together with its extreme hardness and resistance to heat, radiation and chemicals have created a high demand for crystal-clear CVD-diamond plates up to a half an inch (1.27 centimeters) in size for applications in medical, military, aerospace and research industries. Synthetic-diamond optical components are frequently used in the particle acceleration process, laser systems, X-ray focusing equipment, spacecraft windows, telescopes, spectrometers and other analytical instruments and high-powered equipment.

For a long time, man-made diamond lenses have been available for eyeglasses, but the high cost of development has kept them from becoming a mainstream product. Diamond's high index of light refraction makes it well suited to eyeglasses and contact lenses. In addition, CVD-diamond eyeglasses can be made thinner and lighter than traditional lenses, which is especially useful for those with strong prescriptions.

## Unmatched Beauty

When the term diamond is mentioned, most people do not think of ball bearings, machinery, tools, semiconductors or lenses. They picture sparkly gems and diamond jewelry, and these gem-quality diamonds offer different benefits than those with practical applications.

Gem-quality colorless diamonds are brighter than any other natural gemstone because of their extreme hardness, adamantine luster, lack of color and high refractive index. Their fire (the dispersion of light into rainbow colors) is among the highest of any natural transparent gem. This means that even when diamond is colorless, it can display color.

Diamonds are also versatile. They can be worn by anybody with any type or color of clothing and they can enhance any gem they are set alongside. When diamonds are added to colored-gem jewelry, they immediately elevate the perceived value of the other gem and the whole piece.

(opposite page) Diamonds have applications in laser optics. *Andrey Armyagov/Shutterstock*

One of five Type IIb blue diamonds from the Letlapa Tala collection. These five diamonds were found at the Cullinan mine in September 2020 within the space of one week. *Photo © Petra Diamonds*

Sirius Star Octagon cut. *Photo © Dharmanandan Diamonds Pvt. Ltd.*

Diamond's beauty can be unmatched even in its rawest form. Sometimes nature's handiwork alone is enough to inspire awe when viewing a rough diamond, especially if the diamond has a rare color and large size.

It takes millions of years for diamonds to form, and it took centuries for cutters to learn how to bring out their full potential beauty. Today diamonds have a wide range of creative cutting styles, such as the Sirius Star Octagon cut shown above. This innovative cut was conceived by designer Mike Botha and is manufactured under an exclusive global agreement by Dharmanandan Diamonds.

Diamonds also inspire jewelers to create magnificent jewelry pieces, some of which take years to make.

A 1.44-carat heart-shaped Fancy Light pink diamond takes center stage in a dazzling cluster design. The 16 pear-shaped pink diamonds are all from the famed Argyle mine in Western Australia and are laser inscribed. Pink diamonds are among the most beloved and rare of all natural-color diamonds. Finding this many matching pink diamonds is a real challenge for jewelers. *Ring and photo courtesy of Brian Denney of Gems of Note*

(opposite page) The Majestic Necklace is a masterpiece of more than 110 carats of diamonds, with an 8.47-carat pear-shaped D-color $VVS_1$ diamond at the center. The complexity of this piece rests in the flawless matching and attention to detail required to assemble the necklace's 207 diamonds in the handmade platinum mounting. Each ribbon dangling from the length of the necklace is meticulously measured for an impeccably symmetrical silhouette. The diamonds range in color from D to F and in clarity from $VVS_1$ to $VS_2$. The necklace is accompanied by 29 GIA diamond grading reports. *Necklace and photo courtesy of Dehres Ltd.*

## Emotional Significance

Diamonds last forever; they don't fade, dry out or get damaged by chemicals, so they are a perfect heirloom that can be passed down through generations. They can be reset in modern jewelry as a way to honor your loved ones and remind you of the stories associated with those diamonds. Using heirloom diamonds for custom pieces not only makes the jewelry more personal and meaningful but also more affordable.

The highlight of a marriage proposal is the acceptance of an engagement ring, which is typically set with a diamond. Afterward the announcement of the engagement is often made by publicly showing the ring. Some naysayers claim this "tradition" is an invention of modern marketing, but this is not true. Diamonds have been a part of proposals and engagements in Europe for centuries, and diamond rings are now symbols of commitment and enduring love all over the world. Advertising combined with an increased availability of diamonds simply made diamond engagement rings more popular.

The emotional benefits of diamonds have also increased demand for them. A gift of diamond jewelry can make one feel loved and happy. Just looking at that gift afterward can bring back pleasant memories and make one momentarily forget their everyday worries and frustrations. A widow's grief can be interrupted by memories of some of the happiest moments of her life as she looks down at her diamond engagement ring.

The average person is likely unaware of the extent to which diamonds have improved their life. The agricultural, construction, manufacturing, travel, telecommunications and electronics industries all rely on diamonds, as does the medical community. It is an understatement to say that diamonds are just a girl's best friend. They are everybody's friend, thanks to their remarkable intrinsic properties and benefits and the beauty and joy they bring to our lives.

A diamond engagement ring by Mark Schneider Design. *Photo courtesy of Mark Schneider Design*

(opposite page) Heirloom diamond jewelry set with diamonds from rings that were handed down by loved ones but seldom worn. Besides being more modern and stylish, a new heirloom family ring or pendant can be a sentimental remembrance of multiple family members. This heirloom jewelry is by Barbara Westwood, an artist-designer who expresses her innermost thoughts and creativity in the three-dimensional art form of jewelry. *Photos by Sky Hall*

# Glossary

**AGS**: American Gem Society

**Alluvial deposit**: A secondary deposit where rough gemstones accumulate in the sand and gravel of dried-up riverbeds, rivers, streams and ocean shores as the result of erosion and weathering of host rock.

**Alluvium**: The silt, sand and gravel of deltas, dried-up riverbeds and active rivers and streams.

**Antique jewelry**: Any jewelry that is 100 or more years old.

**Bezel facet**: Any of the four-sided, kite-shaped facets on the crown of a round brilliant–cut diamond.

**Blemish**: A flaw on the surface of a gemstone such as a scratch, pit or abrasion.

**Brightness**: The actual and/or perceived amount of light return. In the case of a diamond, it is the combined effect of its surface and internal white light reflections.

**Brilliance**: Brightness with an attractive distribution of dark and bright areas that display good contrast. (This is based on the American Gem Society's brilliance definition of "brightness + positive contrast.")

**Brilliant cut**: The most common style of diamond cutting. The standard brilliant cut consists of 32 facets plus a table above the girdle and 24 facets plus a culet below the girdle. Other shapes, besides round, can be faceted as brilliant cuts.

**Brillianteering**: The final phase of diamond cutting, when the cutter places the star facets, upper half facets and lower half facets on the diamond.

**Briolette**: A gemstone with a teardrop shape, a circular cross section and brilliant-style facets or, occasionally, rectangular step-cut-style facets.

**Bruting**: The process in diamond cutting in which the girdle is shaped, often using another diamond.

**Canary colored**: Yellow

**Carat**: A unit of weight equaling one-fifth of a gram that is used for weighing gemstones. Do not confuse carat, the unit of weight for gemstones, with karat, the unit for measuring gold purity. These two words originate from the same source — the Italian *carato* and the Greek *karation*, which both mean "fruit of the carob tree." In ancient times, carob beans were used as counterweights when weighing gems and gold. Outside the U.S., the word karat is often spelled "carat," particularly in the U.K. and Commonwealth countries.

**Carbonado**: Polycrystalline diamond consisting of diamond, graphite and amorphous carbon. It is black, opaque, porous and tougher than single crystal diamond.

**Champagne colored**: Light brown

**Circa dating**: An approximate date of origin for a jewelry piece. It covers a 10-year window on either side of the date.

**Circular brilliant**: The GIA term for a 58-facet round brilliant-cut gemstone that has lower half facet lengths less than or equal to 65 percent, star facet lengths less than or equal to 50 percent and a medium culet size or larger. The diamond trade generally refers to this type of diamond as a **transitional cut**.

**Slice cut**: A cutting style that uses a laser to slice rough gem material typically about 1.5 to 2.5 millimeters (about 0.06 to 0.10 inches) thick. The surface of a slice may be entirely flat and smooth, or it might have some very low-angled large facets to create a sparkle effect from light reflected from its top surfaces.

**Splittable**: Diamond rough that can be divided into small but valuable segments by lasering or cleaving.

**Step cut**: A cutting style with rows of facets that resemble the steps of a staircase when viewed from above and below. The facets are usually four sided, elongated and parallel to the girdle.

**Symmetry**: A grading term for the exactness of shape and placement of facets. It is one of two subcategories of finish.

**Synthetic diamond**: A diamond made in a laboratory that has the same chemical composition and crystal structure as natural diamond. Their physical and optical properties are almost the same as those of natural diamonds. See chapter 7 for more information.

**Table**: The large, flat top facet; it is octagonal on a round brilliant.

**Table cut**: A cutting style consisting of an octahedral shape with its top point cut away, creating a square flat-top table facet.

**Tetrahedron**: A group of five bonded carbon atoms, with one carbon atom in the center and the other four atoms surrounding it. Tetrahedrons, in turn, combine to form a unit cell, and many unit cells together form a diamond crystal.

**Toughness**: The resistance of a gemstone to breaking, chipping or cracking.

**Transitional cut**: A cutting style often used in the 1930s and '40s that features the overall faceting of a modern brilliant cut but with a slightly open culet and a bit of faceting that resembles an old European cut.

**Translucent**: Allowing light to pass through partially, like frosted glass.

**Transparency**: The degree to which a gemstone transmits light. In other words, the degree to which the stone is clear, cloudy or nearly opaque.

**Transparent**: Crystal clear; objects seen through transparent gemstones look sharp and distinct.

**Treatment**: The standard term used in gemological literature for any process done to improve the appearance of a gemstone (not including cutting and cleaning).

**Type I diamond**: A diamond with nitrogen in its chemical structure. This affects the physical and optical properties of the diamond.

**Type II diamond**: A diamond without significant nitrogen in its structure.

**Ultraviolet light**: Radiation with wavelengths from 10 nanometers to 400 nanometers. It is used to help identify gemstones.

**Zoning**: Alternating sections of color in a gemstone.

## WEBSITES

Alrosa.ru
AntiqueJewel.com
AuctionMarketResource.com
Canada.DeBeersGroup.com
DDMines.com
DeBeersGroup.com
DebmarineNamibia.com
Debswana.com
GemDiamonds.com
GemologyOnline.com
Geology.com
GeorgianJewelry.com
GIA.edu
HA.com
IdexOnline.com
LangAntiques.com/university
Langerman-Diamonds.com/encyclopedia
Lucapa.com.au
LucaraDiamond.com
Mindat.org
Mining.com
Mining-Technology.com
Namdeb.com
NaturalDiamonds.com
NRCan.gc.ca
PetraDiamonds.com
RioTinto.com
RioZim.co.zw
USGS.gov

# Acknowledgments

I'd like to thank the following people and companies for their contributions to this book:

The staff at Firefly Books responsible for producing this beautiful, high-quality book: Lionel Koffler, president; Julie Takasaki, managing editor; Ronnie Shuker, copy editor; George A. Walker, map and diamond-cut illustrator; and Hartley Millson, designer. I have greatly appreciated the opportunity to create this book with such dedicated and talented professionals.

Nancy Almeida, Diana Asonova, Marianna Bowes, Paul Cassarino, Michael Cowing, Branko Deljanin, Al Gilbertson, Hitesh Goti, Terry Kruger, Gary Megal, Daniel Nyfeler, Lisa Stockhammer-Mial and Sandy Ray. They have provided information and made valuable suggestions, corrections and comments. They are not responsible for any possible errors, nor do they necessarily endorse the material contained in this book.

The teachers at GIA. They helped me obtain the technical background needed to write a book such as this. Their dedication and assistance extend well beyond class hours.

Abe Mor Diamond Cutters, ACA Gem Lab, Adin Fine Antique Jewelry, Alrosa, The Arkenstone, Asian Star Group, John Atencio, Laurent Cartier, Paul Cassarino, John Chapman, Michael Cowing, Crater of Diamonds State Park, Paula Crevoshay, De Beers Canada, De Beers GemFair, Dehres Ltd., Branko Deljanin, Brian Denney, Dharmanandan Diamonds Pvt. Ltd., Diamond Foundry, Element 6 Group, Gemological Institute of America, Gems of Note, Gems by Pancis, GeorgianJewelry.com, Jim Grahl, Gübelin Gem Lab, Sky Hall, Barbara Heinrich, Heritage Auctions, Herbert Horowitz, Hubert Jewelry, Peter Indorf, KT Diamond Jewelers, Anup Jogani, J. Landau Inc., Lang Antiques, Glenn Lehrer, Lightbox Jewelry, Lotus Colors Inc., Lucapa Diamond Company Limited, Lucara Diamond, Meylor Global, Abe Mor Diamond Cutters, Moussaieff Jewellers, Namdar Diamonds, Namdeb Diamond Corporation, New World Diamonds, Pala International, Petra Diamonds, Todd Reed, Sandvik Additive Manufacturing, Mark Schneider Design, Dusan Simic, Single Stone, Smiling Rocks, Smithsonian, State Parks of Arkansas, The Three Graces Jewelry, Vrai and Barbara Westwood. Photos and/or diagrams from them have been reproduced in this book.

Ernie and Regina Goldberger of the Josam Diamond Trading Corporation. I will forever be grateful to them for hiring me to sort and grade their diamonds and oversee their jewelry production. This book could never have been written without the experience and knowledge I gained from working with them.

# Index

*Page numbers in italics refer to illustrations.*